DRUG ABUSE

Nonmedical Use of Dependence-Producing Drugs

ADVANCES IN EXPERIMENTAL MEDICINE AND BIOLOGY

Volume 1
THE RETICULOENDOTHELIAL SYSTEM AND ATHEROSCLEROSIS
Edited by N. R. Di Luzio and R. Paoletti • 1967

Volume 2
PHARMACOLOGY OF HORMONAL POLYPEPTIDES AND PROTEINS
Edited by N. Back, L. Martini, and R. Paoletti • 1968

Volume 3
GERM-FREE BIOLOGY: Experimental and Clinical Aspects
Edited by E. A. Mirand and N. Back • 1969

Volume 4
DRUGS AFFECTING LIPID METABOLISM
Edited by W. L. Holmes, L. A. Carlson, and R. Paoletti • 1969

Volume 5
LYMPHATIC TISSUE AND GERMINAL CENTERS IN IMMUNE RESPONSE
Edited by L. Fiore-Donati and M. G. Hanna, Jr. • 1969

Volume 6
RED CELL METABOLISM AND FUNCTION
Edited by George J. Brewer • 1970

Volume 7
SURFACE CHEMISTRY OF BIOLOGICAL SYSTEMS
Edited by Martin Blank • 1970

Volume 8
BRADYKININ AND RELATED KININS: Cardiovascular, Biochemical, and Neural Actions
Edited by F. Sicuteri, M. Rocha e Silva, and N. Back • 1970

Volume 9
SHOCK: Biochemical, Pharmacological, and Clinical Aspects
Edited by A. Bertelli and N. Back • 1970

Volume 10
THE HUMAN TESTIS
Edited by E. Rosemberg and C. A. Paulsen • 1970

Volume 11
MUSCLE METABOLISM DURING EXERCISE
Edited by B. Pernow and B. Saltin • 1971

Volume 12
MORPHOLOGICAL AND FUNCTIONAL ASPECTS OF IMMUNITY
Edited by K. Lindahl-Kiessling, G. Alm, and M. G. Hanna, Jr. • 1971

Volume 13
CHEMISTRY AND BRAIN DEVELOPMENT
Edited by R. Paoletti and A. N. Davison • 1971

Volume 14
MEMBRANE-BOUND ENZYMES
Edited by G. Porcellati and F. di Jeso • 1971

Volume 15
THE RETICULOENDOTHELIAL SYSTEM AND IMMUNE PHENOMENA
Edited by N. R. Di Luzio and K. Flemming • 1971

Volume 16A
THE ARTERY AND THE PROCESS OF ARTERIOSCLEROSIS: Pathogenesis
Edited by Stewart Wolf • 1971

Volume 16B
THE ARTERY AND THE PROCESS OF ARTERIOSCLEROSIS: Measurement and Modification
Edited by Stewart Wolf • 1971

Volume 17
CONTROL OF RENIN SECRETION
Edited by Tatiana A. Assaykeen • 1972

Volume 18
THE DYNAMICS OF MERISTEM CELL POPULATIONS
Edited by Morton W. Miller and Charles C. Kuehnert • 1972

Volume 19
SPHINGOLIPIDS, SPHINGOLIPIDOSES AND ALLIED DISORDERS
Edited by Bruno W. Volk and Stanley M. Aronson • 1972

Volume 20
DRUG ABUSE: Nonmedical Use of Dependence-Producing Drugs
Edited by Simon Btesh • 1972

Volume 21
VASOACTIVE POLYPEPTIDES
Edited by N. Back and F. Sicuteri • 1972

DRUG ABUSE

Nonmedical Use of Dependence-Producing Drugs

Proceedings of the Sixth Round Table Conference organized by the Council for International Organizations of Medical Sciences with the participation of the United Nations Educational, Scientific, and Cultural Organization and the World Health Organization, Geneva, Switzerland, October 20-21, 1971

Edited by

Dr. Simon Btesh

Executive Secretary, CIOMS

PLENUM PRESS • NEW YORK - LONDON • 1972

Library of Congress Catalog Card Number 76-190227
ISBN 0-306-39020-5

A Division of Plenum Publishing Corporation
227 West 17th Street, New York, N.Y. 10011

United Kingdom edition published by Plenum Press, London
A Division of Plenum Publishing Company, Ltd.
Davis House (4th Floor), 8 Scrubs Lane, Harlesden, London, NW10 6SE, England

Printed in the United States of America

Chairman:	A. GELLHORN
Moderators:	J. JACOB D. X. FREEDMAN K. EVANG
Speakers:	R. ANDREASSON K. BASH P. BEEDLE D. C. CAMERON T. L. CHRUSCIEL P. DENIKER K. C. DUBE G. EDWARDS A. HUSAIN H. ISBELL P. KIELHOLZ V. KUSEVIC H. C. LEONG J. MABILEAU J. MENDELSON J. de MOERLOOSE B. REXED S. SHOHAM R. C. SMITH L. STEINIG
Secretary:	S. BTESH

NOTE ON CIOMS

The Council for International Organizations of Medical Sciences was established in 1949 under the joint auspices of the World Health Organization and UNESCO. The Council may be regarded as a federation of non-governmental organizations dedicated to the progress in the medical sciences. There are more than fifty "international members" of the Council covering not only the basic disciplines of medicine but also the various clinical disciplines. In 1961 a category of "national members" was created to enable a single national body such as a national academy of medicine or sciences or a national research council to be admitted as a member.

The work of the Council includes the following: the coordination of international congresses; the convening of international meetings, especially round table conferences on the main problems of modern medicine; the establishment of an International Nomenclature of Diseases in four languages (English, French, Russian, Spanish).

The General Assembly of CIOMS, constituted of delegates of its member organizations, meets every three years and appoints an Executive Committee, which is in charge of administration in the interval between meetings of the General Assembly. The Secretariat is directed by the Executive Secretary.

The statutes of the Council have been altered in 1967, especially in relation to the aims of the organization, which have been broadened with a view to collaboration with the United Nations and its specialized agencies, to encouragement of international activities in the field of medical sciences, and to the service in general of the scientific interests of the medical community of the world.

RECENT CIOMS PUBLICATIONS

BIOMEDICAL SCIENCE and the DILEMMA of HUMAN EXPERIMENTATION. Round Table Conference, 1967, 110 pages, 13x21 cm. Published by Desoer, Liège, Belgium.

The PLANNING of INTERNATIONAL MEDICAL MEETINGS, 1967, 115 pages, 20x25 cm. Published by the Union of International Associations, 1, rue aux Laines, Brussels 1.

PROCEEDINGS of the SYMPOSIUM on SCIENCE POLICY and BIOMEDICAL RESEARCH. 1968, 76 pages, 21x27 cm. Published by UNESCO in the Series "Science policy studies and documents" No. 16.

HEART TRANSPLANTATION Round Table Conference, 1968 53 pages, 13x21 cm. Published by Desoer, Liège, Belgium.

EVALUATION of DRUGS: WHOSE RESPONSIBILITY? Round Table Conference, 1969, 101 pages, 13x21 cm. Published by Desoer, Liège, Belgium.

MEDICAL RESEARCH: Priorities and Responsibilities. Round Table Conference, 1969, 156 pages, 13x21 cm. Published by the World Health Organization, Geneva.

The TRAINING of RESEARCH WORKERS. Round Table Conference, 1970, In Press.

Obtainable at the CIOMS Offices :
1 Rue Miollis (UNESCO), Paris, France.
Avenue Appia, (WHO), Geneva, Switzerland

CONTENTS

FIRST SESSION - INTRODUCTORY

Chairman - Dr. Alfred Gellhorn

President of CIOMS

DR. GELLHORN: The round table conference on "Non-medical Use of Dependence Producing Drugs - Current Problems and Approaches" is now formally in session. I wish to welcome the large representation from international biomedical societies, from the academies of science and medicine of many nations and the participants of the round table. Before calling upon the distinguished representative of UNESCO and the distinguished representative of WHO for their opening remarks, I should like to set the stage for this conference by recording briefly the background of the Council of Organizations of the Medical Sciences.

The genetic constitution of CIOMS is noteworthy in that it was created as a joint venture of WHO and UNESCO in 1949. In this immediate post-world war II period the principal assignment of CIOMS* was to re-establish international scientific and medical communication and facilitate this by aiding in the organization of international congresses. Although CIOMS continues to fulfill this function and to co-ordinate the international meetings through a detailed calendar listing which is widely distributed, it has undertaken other activities including the development of symposia and round table conferences such as this one.

* At that time called the Permanent Committee for Coordination of International Congresses and changed to the Council for International Organizations of the Medical Sciences (CIOMS) in 1952.

A characteristic of the conferences, and one which shows the influence of CIOMS parents, UNESCO and WHO, is that they are organized to bring out the interrelationships of the science and the art of medicine, the ethical as well as the technical, the pragmatic and the philosophical, the role of the social sciences as well as the medical sciences in the important issues of health today. This can be seen in the published reports of earlier conferences such as Medical Research-Priorities and Responsibilities or The Training of Research Workers, and the present conference demonstrates dramatically the interdependencies of the medical and social sciences. In the succeeding year conferences on The Social and Ethical Implications of Progress in Biology and Medicine and New Trends in Medical Education and Their Implications for Family Health also will speak to the science and to the humanity of the topics.

The Council for International Organizations of the Medical Sciences now has fifty-three international members which represent essentially all of the international biomedical associations. Additionally there are seventeen national members representing the national academies of medicine or science of their respective countries. This broad international membership, together with the close continuing relationship maintained with UNESCO and particularly with WHO, provides the Council with strength and purpose.

I now call on Mr. A. Husain for a message from UNESCO.

MR. HUSAIN: Mr. President, in the name of the Director General of UNESCO, I wish to welcome the participants in this Round Table Conference which has been convened by the Council of International Organization of Medical Sciences. UNESCO is interested in this Conference for two particular reasons. Firstly of its continuing working relationship with the Council itself. Secondly because of its interest in the development of the Drug Education Programe. I am sure Mr. Chairman, that the deliberations of this Conference will be useful not only for the participants but also for the agencies and organizations they represent.

DR. GELLHORN: I thank you Mr. Husain for your expression of interest and concern of UNESCO for the proceedings of this conference.

CIOMS is particularly indebted to WHO for its continuing financial assistance and we are all indebted to WHO for its courtesy in allowing us the use of the magnificent chamber of the WHO Council in which these meetings are to be held. Dr. T.A. Lambo, a distinguished psychiatrist, former Dean of the School of Medicine at Ibadan, and onetime member of the Executive Committee of CIOMS is now the Assistant Deputy Director General of the World Health Organization and it is a particular pleasure and privilege to recognize him. Dr. Lambo.

DR. LAMBO: Mr. President, Members of the Executive Committee, Distinguished Members of CIOMS, Ladies and Gentlemen. It is my great pleasure and privilege to welcome all of you here this morning on behalf of the Director General, Dr. Candau, who is unavoidably absent from Geneva at this time. CIOMS constitutes one of the most important organizations with which WHO collaborates and derives a lot of benefit and insight into some of the contemporary health problems of our time. We have taken a great deal of pride in the endeavours and the activities of CIOMS, not because of the genetic constitution and relationship it bears to us, not because we are one of the co-sponsors of these very important organizations, but because of the vitality with which it pursues some of its deliberations. The Round Table Conference, or conferences, which CIOMS has held have been of great scientific benefit to the entire world. I believe that this is the sixth one and I remember the first one on Biochemical Science and the Dilemma of Human Experimentation in which I was privileged to take part. I did not only enjoy a sheer intellectual atmosphere, because I remember very well that it was Professor Hersch, Professor of Philosophy of the University of Geneva, who then was at UNESCO, who took the Chair, but I also enjoyed the depth of the scientific enquiry and the penetration. The present subject, Non-Medical Use of Dependence Producing Drugs - Current Problems and Approaches is of great interest not only to WHO but to the entire world at large. And this is evidenced by the quality and number of people that are present here this morning. Mr. President it is with great pleasure that I welcome all of you to WHO and we are extremely proud indeed that this Round Table Conference is taking place here in Geneva.

DR. GELLHORN: Thank you, Dr. Lambo.

The subject of drug abuse is one which has been widely and frequently discussed. The conference of the next two days differs from others in the broad international representation and the interdisciplinary approach to the issue. Thus the fields of education, sociology, law, epidemiology, pharmacology, public health administration, psychiatry, surgery and medicine are effectively represented among the round table participants, and the exchange of information and opinions between these experts will, I believe, bring forth fresh ideas and approaches to an important problem of our time. For the concerned general audience I expect us to learn:

1. A definition of the problem of drug abuse and a clearer idea of its extent.

2. A knowledge about the drugs currently being abused which is at least equal to that of the users.

3. What can be done about prevention?

4. What should the physician know about treatment?

I should like now to call on Dr. Fattorusso, Director of the Division of Pharmacology and Toxicology of the World Health Organization.

DR. VITTORIO FATTORUSSO: I must first thank you for giving me the opportunity to say a few words about the World Health Organization's programme in the field of drug dependence or, in a broader sense, of what is commonly called drug abuse. I shall confine myself to a few very general remarks.

I must explain that, since its inception, the World Health Organization has exercised functions conferred upon it under various international agreements which have instituted a system for control of narcotic drugs. Over the years there has been a succession of international agreements concluded on narcotics control and the World Health Organization has exercised, and still exercises, functions under the 1931 Convention for Limiting the Manufacture and Regulating the Distribution of Narcotic Drugs, as amended by the Lake Success Protocol of 1946, and the Single Convention on Narcotic Durgs, signed in New York in 1961. The purpose of this latter convention was to consolidate the previous agreements; it defines the term "narcotic drug" as designating any of the substances, whether natural or synthetic, in Tables I and II annexed to the Convention, and if we look at these tables we find there, broadly speaking, opium, the opiates, their synthetic substitutes, coca and cocaine, and cannabis. I shall not go into the details of the various international agreements on narcotics, but I should like to stress that, in their application, it is the World Health Organization which, on its own initiative or in response to a notification from the Secretary-General of the United Nations, recommends to the Commission on Narcotic Drugs of the Economic and Social Council the appropriate measures for international control or the amendment of the existing provisions.

I made a point of referring to the international agreements on narcotic drugs because the exercise of the functions conferred upon the World Health Organization by these treaties is in fact the oldest part of its programme in the drug field. For the exercise of these functions, the First World Health Assembly had already agreed in 1948 to the establishment of a WHO Expert Committee; since the, this Committee has met eighteen times and has not only issued opinions and recommendations on measures for control of narcotics, but has looked into a whole range of other problems. The Expert Committee and the World Health Assembly have for years been stressing the need to place under international control other substances which are not covered by the agreements on narcotic drugs but which give rise to abuses. These substances include, in particular, amphetamines, barbiturates, hallucinogens and tranquillizers that produce dependence - in a word, what are nowadays known as psychotropic substances. These recommendations of the WHO Expert Committee provided the basis for drawing up the list of substances to be placed under international control by the new Convention on psychotropic substances, which was signed in Vienna at the beginning of this year and will come into force when forty countries have ratified it. In the case of the psychotropic substances covered by the new Convention, as in that of narcotic drugs, it will be for the World Health Organization to evaluate them and make recommendations on the appropriate control measures.

Of course, the interests and activities of the Organization are not restricted to the evaluation of narcotics and psychotropic substances. The World Health Assembly has recommended that the Organization should encourage and assist the development of preventive, treatment and rehabilitation programmes; it has also approved an expansion of the programme, including in particular collection and exchange of information, analysis of all the medical, social, cultural and economic factors which contribute to drug dependence, and the organization of research programmes. It is on this very aspect of research that my two colleagues, Dr. Cameron and Dr. Chrusciel, will have an opportunity to give you a few details this afternoon and tomorrow.

In this multidisciplinary approach to the problems posed by drugs, the World Health Organization is collaborating with other agencies and bodies of the United Nations system. But in this endeavour it is also very important that the non-governmental organizations should make their contribution. The Council of International Organizations of Medical Sciences, which represents a wide sector of the world medical and scientific community, is particularly well placed to tackle this problem from the multidisciplinary angle and by virtue of its traditions constitutes a forum for free discussion. We are accordingly gratified, Mr. Chairman, at the initiative it has taken in organizing this sixth Round Table and we shall follow your discussions with great interest. Allow me, in my turn, to wish you the greatest possible success.

INTRODUCTORY ADDRESS

DRUG DEPENDENCE - PUTTING HISTORY ON THE AGENDA

Dr. Griffith Edwards

Addiction Research Unit

Institute of Psychiatry, London

DR. EDWARDS: Today's problems make such urgent demand that we sometimes overlook the extraordinary stretch of their history. We feel frustrated if we cannot put matters right at a stroke while forgetting that the world has coexisted with rather similar problems perhaps for a thousand years. The Arabs took the poppy to China in the 8th Century (EDKINS, 1894).

The objection will be heard that our own times are too fraught usefully to allow diversion of energies to historical study. We may plead that we have no time to tarry, that we must build the next hospital, write the research protocol, draft the new legislation. But the historian (if we pause to listen to him) forces questions such as these - what scientific truths or old wives tales, what culture bound or generation determined illusions, accidents, vanities are the actual ideas dictating all this mass of action in which we propose so hurriedly to engage?

The purpose of the present short paper cannot of course be to treat the vast and complex business of the whole history of man's relationship to psychoactive substances - such an endeavour would be quite outside proper competence. The aim can only be to identify and list out certain questions for an historical agenda - and this list is incomplete.

Perhaps the first question for such an imaginary agenda should be as to the history of official information. What sort of data have decision-makers had available to them when shaping the nature of a society's response to a substance? What data did the Chinese have on tobacco when roughly 300 years ago they banned its use? What data will then be needed before a country today puts a compulsory warning on a packet of cigarettes? Absence of data may be as important as its presence.

In addressing himself to this information question the historian might find the study of committees and commissions provided him with interesting material. He might, for instance, wish to determine the nature of the societal forces which could be identified as having caused each committee to be set up - The Royal Commission on Opium 1894-1895 could provide a case-study (OPIUM COMMISSION, 1894). Parliament was responding to live and well organized lobbying by the Anti-Opium Society. During the session of 1891, a total of 3, 352 petitions were presented at Westminster and these bore a total of 192, 106 signatures. Something of the fervour is preserved in a memorandum which the President of the Anti-Opium Society submitted to the Commission:-

> "The cancer that has been eating into the vitals of Chinese life will be torn out; and once again, in many places, a population morally and physically vigorous, will be ready to 'receive the engrafted Word'... the constitution of railways will be encouraged: a free interchange will be established of the products of British and Chinese industry..."

Evangelism and business interest seemed happily as so often do to coincide.

Something is also to be learnt from the covert implications of the actual wording of the terms of reference which are given to a committee. When in 1834 Parliament set up a group to look at the problem of drunkenness (SELECT COMMITTEE, 1834) its terms of reference were:-

> "... to inquire into the extent, causes and consequences of the prevailing vice of intoxication among the labouring classes of the United Kingdom, in order to ascertain whether any legislative measures can be devised to prevent the further spread of so great a national evil".

The covert message implied by such a brief is that those who framed it saw themselves as concerned with a "vice" and an "evil", that they were concerned with prevention rather than treatment, and that they believed in the likely effectiveness of legislative remedies. Their brief also admits the belief that the problem was peculiarly one of a particular social class. Several committees and over one hundred years later the same country found itself very perplexed by the problem of public drunkenness, and the Home Office set up a Working Party:-

> "...to consider the treatment, within the penal system, of offenders who habitually commit offences involving drunkenness..." (HOME OFFICE 1971).

How does this second message compare with its nineteenth century forebear? The message is now ambiguous as to whether the concern is with vice or with illness, for the working party's task is to consider treatment (implying illness) within the penal system (implying vice). In that the focus is on the chronic drunkenness offender rather than on alcoholism in general, the propriety of a class focus is again implicit. Gone however is any concern with prevention - the emphasis is on treatment.

Membership and brief are closely linked. The two Brain Committees (MINISTRY OF HEALTH 1961, 1965) were with the exception of a pharmacologist, medical in constitution - membership of the Wootton Committee on cannabis (HOME OFFICE 1968), with its wide representation of interest and social science chairman, shows more certainly than any ministerial statement the abrupt change toward official acceptance of the notion that the drug problem was not just the province of the doctor.

The general trend as regards witnesses - there are exceptions - has been similar to that which can be seen with committee membership. The layman has been devalued and the representative of the profession, of the special interest, is the person who is expected to speak. The Opium Commission (OPIUM COMMISSION 1895) saw 723 witnesses and among these, landowners and tenants, planters, merchants, mill-owners and bankers out-numbered any of the professions.

Committees have at times also gathered evidence from

other sources than witness and written submission. The La Guardia Committee (MAYOR'S COMMITTEE ON MARIHUANA, 1944) commissioned its own research. The Le Dain Commission (1970) has provided a possibly unique example of imagination, thoroughness and energy in data collection...It must be evident by now that even the business of listing the main elements which should be studied in any analysis of drug and alcohol committees, of identifying the covert messages, would if properly conducted by a very extensive task indeed. Such analysis must be left as unfinished business and we may turn then very briefly to consideration of what other items might be put on the historical agenda. Perhaps it would add to the illusion of the agenda if further items were in traditional committee style actually listed out and numbered.

Item 2. HISTORY OF THE RESEARCH WORKER'S ROLE

Hawks (1971) has recently presented a provocative essay on the place (or lack of place) of social research as a determinant of social policy. The social scientist, as we have seen him develop in this century, has "disdained any sense of public accountability" and has hence "forfeited any claim to a public audience". Hawks quotes Cherns (1968) as saying:-

> "I have not been able to trace in practice one example of a study carried out in a university or research institute or anywhere else which resulted in direct application except where the researcher had become involved in following through his studies into application".

How would such remarks apply to the history of the relationship between research and action in the alcohol and drug field? Who historically has here determined research priorities?

Item 3. THE HISTORY OF THE USER'S ALIENATION

In different societies the user of a particular substance has been at a greater or lesser distance from the decision-makers, and probably the greater the distance the greater the fantasies. Adolescent subculture, whether or not drug using, is likely to be at a great social distance from the sub-culture of the senior civil servant: drug taking did not look so strange, was not so worrying, did not excite such reaction when the drug-taker was a middle-aged member of the middle classes. Writing in a medical journal in 1882, Dr. C. R. Francis described the case of "a friend now in his 6th. decade who had taken opium for about 30 years:-

> "Yielding to the popular prejudice against opium eating, Mr. A. has repeatedly tried to break it off... Doubtless he would succeed in time, as others have done, but cui bono?"

To what extent is the history of alienation a history of the inevitable, and to what extent is it the history of anxious social reaction gone wrong? How often has legislation blown bridges?

Item 4. THE HISTORY OF LEGISLATIVE EFFORT

Society's belief that laws must regulate the use of alcohol (and drugs) is ancient, and it would be fascinating to examine the historical evidence that can be culled for the success or failure of legislation. The major finding might well be of society's general tendency to believe that when harsh measures are ineffective, they are not discredited - history might show that it was the rather exceptional instance in which failure of harshness led to other than greater harshness. A barrister writing on the management of Inebriates Retreats (WOOD RENTON, 1896) suggested that new powers were needed, above those existing which allowed for punishment of the inebriate who would not work - he wanted a fine of £5, or 7-day imprisonment, for the man who would not take exercise.

Item 5. HISTORY OF COMPULSORY TREATMENT

Compulsion is likely to be favoured at times when the focus has swung toward seeing the user as the problem rather than the society which generates use as the problem. Compulsion is perhaps also more likely to be in favour when the user is alienated (skid row alcoholic, adolescent drug user) than when the decision-makers can more or less identify with him (middle-class alcoholic, "professional" drug addict). It will be more favoured when society is more anxious. The first Brain Report (MINISTRY OF HEALTH, 1961) found against compulsion while four years later (MINISTRY OF HEALTH, 1965) the second report recommended introduction of compulsory treatment -

during those four years no further evidence had accrued of treatment being effective nor of compulsion aiding treatment - the typical narcotic user had however changed in type, and there were more users.

Item 6. HISTORY OF THE ABSOLUTE SOLUTION

An analysis might be made of the nature of those forces which have proposed an absolute solution to society's drug or alcohol use - which have believed in the radical cure of problems rather than in the feasibility of coexistence. The classic example of root-and-branch which most readily springs to mind, is perhaps the American experiment with alcohol prohibition, but a close analysis might indeed suggest that the Western cast of mind has in the last decades very generally been actuated by absolutism in its dealings with the non-medical use of nearly every psychoactive substance other than alcohol. Absolutism may be the potent underlying fantasy even when it is not the outspoken manifesto. The history of the clash between the absolutists and their opponents, the conditions in which one or the other camp has won the victory, might be determined. The case against absolutism has seldom been more passionately put than by Batten (1894), when he was attacking the position of the Anti-Opium Society:-

> "In conclusion, I would say to the Society for the Suppression of the Opium Trade - You may make this a party question; you may win votes relying on the uninstructed philanthropy of your fellow countrymen...you may cover India with an army of preventive officers to watch another army of smugglers, created to satisfy a natural demand of which you have suppressed the legitimate supply; you may harass the people by personal searches and domiciliary visits; you may thus create discontent among our native subjects... you may drive the consumers of opium to alcohol and hemp... but you will never persuade the Chinese to follow in your footsteps and abandon the ever increasing culture of the poppy....."

Was Batten right in his arguments, or with the hindsight of history do we see him resisting absolutism when absolutism would have been proper?

Item 7. THE HISTORY OF DRUGS AND ALCOHOL PROBLEMS AS POLITICAL ISSUE

An historical analysis might be made of the circumstances which have at times caused drug and alcohol questions to have become entangled with politics, and the happy or unhappy consequences of such politicization. In examining this matter account would certainly have to be taken of the extent to which governments have profited from sale or tax, and of the processes which lead to the acceptance or rejection of such involvement. The annual cash value of the Indian opium crop was approximately £ 13, 100, 000 (BATTEN, 1894). When international politics or international profit effect the drug issue, matters become very complex indeed.

Item 8. THE HISTORY OF MYTHS

A rather more light-hearted essay might perhaps be put together on the history of various myths regarding alcohol or specific drugs, which have entered into currency. The Indian asylum statistics on cannabis-induced mental illness might provide an example. The unreliability of these figures was uncovered by the Indian Hemp Drugs Commission (1894) - the members of that Commission had a habit of sharp questioning:-

> "*First*, as to procedure, Dr. Crombie told the Commission that the descriptive roll is "by no means trustworthy" but that 'subsequent discoveries' were made in the asylum by examination of friends or of the lunatic on recovery and by other means, and that the entries in the asylum books would be altered accordingly. 'This' he said 'was my practice'. The Commission had found no trace of this practice in the asylum records; and they showed Dr. Crombie and asked him to point out such alterations. He then withdrew his statement... Dr. Crombie's procedure then differed in no respect from the most mechanical and unintelligent record of causation in any asylum in India....."

And yet Lewin (1931) in his much respected classic referred to "information" which had been forthcoming from "the lunatic asylums in India", without any reference to the Commission's probings. Myths are remarkably immune to the assault of mere fact.

Item 9. HISTORY OF DRUGS AND ALCOHOL AND THE HISTORICAL BACKDROP

No matter how determinedly we insisted that the historian should focus his attention simply on drugs and alcohol, he would sadly have to refuse our demands. He would inevitably tell us that to explain the response of any society at any point in time to drugs and alcohol, one has to look not just at the substances but also at the society. The recent South African report nicely, but no doubt accidentally, makes this point (REPUBLIC OF SOUTH AFRICA, 1970):-

> "At the International Seminar on the Prevention and Treatment of Drug Dependence in Lausanne (8 to 11 June 1970) the platform was, however, unfortunately also used for propaganda for a more permissive attitude towards, and for more freedom for, the individual and especially youth, less interference by the legislature, and more elbow room for social research, views which may certainly be put forward, but left the impression that there were no clearly defined motives present".

Indeed, the historian might well tell us that this very last point on the agenda was the most important one of all. Response to drugs and alcohol are, not discreet and separate from society's evaluation of the nature of responsibility, the nature of liberty and indeed the nature of man.

Every agenda has then of course a traditional entry for "any other business". This, if the Chairman is unwary, is the opportunity for some disruptive person to ask an inopportune question which even at a late moment will wreck the calm of the meeting - the questioner might wish to enquire for instance as to how this actual present meeting had come about, the nature of the information with which it was being fed, its distance from the users of the substances being discussed, this meeting's cultural beliefs as to the place of law and the nature of liberty and its appetite for absolutism, its political affiliations, the myths which may infect its mind, the nature of the world which lies outside its windows.

Any half-way able chairman faced by such a questioner would of course adjourn the meeting - would say that time was up.

REFERENCES

Batten, G.H.M. 1894 Opium Commission. Report of the Royal Commission on Opium. Vol 1. Appendix 1, 133-146 London: H.M.S.O.

Cherns, A.B. (1968) Human Relations 21, 313-325. The use of the social sciences.

Edkins, - (First name not recorded). 1894 Opium Commission. First Report of the Royal Commission on Opium. Historical note on opium and the poppy in China. Appendix II. London: H.M.S.O.

Francis, C.R. (1882). Medical Times and Gazette, 1, 87-89. On the value and use of opium.

Hawks, D.V. (1971). Social research as a determinant of social policy. Paper read to the Second International Institute on the Prevention and Treatment of Drug Dependence, Baden, Austria.

Home Office 1968 Cannabis Report by the Advisory Committee on drug dependence. London: H.M.S.O.

Home Office 1971 Habitual Drunken Offenders: report of the Working Party. London: H.M.S.O.

Indian Hemp Drugs Commission 1894. Report, Vol. 1. Simla: Government Central Printing Office.

Le Dain Commission of Inquiry into the Non-medical use of Drugs. 1970. Interim Report. Canadian Government.

Lewin, L. (1931) Phantastica: narcotic and stimulating drugs. Trans. P.H.A. Wirth, from the second German edition. London: Kegan Paul, Trench and Traubner.

Mayor's Committee on Marihuana (1944). The Marihuana problem in the City of New York. Jacques Caltell Press, Lancaster Pennsylvania.

Ministry of Health 1961. Drug addiction: report of the Interdepartmental Committee. London: H.M.S.O.

Ministry of Health 1965. Drug addiction: the second report of the Interdepartmental Committee. London: H.M.S.O.

Opium Commission 1894. Report of the Royal Commission on Opium Vol 1. London: H.M.S.O.

Opium Commission 1895. Report of the Royal Commission on Opium Vol. 7. London, H.M S.O.

Republic of South Africa 1970. Report of the Committee of Inquiry into the abuse of drugs (Government publication).

Select Committee 1834. Report from the Select Committee on Inquiry into drunkenness. British Parliamentary Papers 559, Vol. 8.

Wood Renton, A. (1896). J Ment. Sci. 42, 85-96. On recent proposals regarding habitual drunkards and other offenders.

DR. GELLHORN: Thankyou Dr. Edwards. What a wonderful way to begin the Conference. Isn't it fortunate that already we have the provocative questioner immediately aboard, and therefore we are prepared not to adjourn the meetings and to accept the challenge that has been presented by Dr. Edwards. As we proceed, perhaps we will examine more carefully some of the issues that he has raised with regard to our own personal cultural background and biases, and to see to what extent they color our discussions and presentations. We shall now open the second session of the conference. I call on Professor Jacob to take the chair.

SECOND SESSION

FACTORS ASSOCIATED WITH THE USE OF DEPENDENCE PRODUCING DRUGS

MODERATOR: Professor Joseph Jacob, Chef du Service de Pharmacologie et de Toxicologie, Institut Pasteur, Paris, France

PROFESSOR JACOB: I should like to begin by thanking you, Dr. Gellhorn, for the honor conferred upon me by inviting me to act as the moderator of this Session. Time is precious and should be reserved for the presentations and discussion of this important subject. I will therefore immediately call on Dr. Daniel X. Freedman, who is a distinguished pharmacologist as well as psychiatrist, to introduce his paper.

NON-PHARMACOLOGICAL FACTORS IN DRUG DEPENDENCE

Daniel X. Freedman, M. D.

Chairman, Department of Psychiatry

University of Chicago

The apparatus with which man engages reality can--in an evolutionary perspective--be seen to have a certain, though wavering skew toward expediency, survival, and the maintenance of sufficient vigilance for a majority to achieve this. It also permits a range of private experiences, as well as a search for comfort and obtunding of pain; these factors can be sustained with various life styles which can be judged as having tolerable or marginal, or thoroughly inadequate survival value. We seem also to be gifted with the capacity to accentuate and elaborate these capacities and vulnerabilities for subjectivity with a variety of tools and practices; we can implement megalomanic as well as retreatist or reality goals.

In enhancing our purposes with chemicals, we seem, nevertheless, to be intrinsically limited in the extent to which we can endure chemical pleasures. I have often wondered whether the fact that drugs rarely sustain their initial allure and promise was a mischievous and malevolent trick of nature. Whatever the problems of design, there do appear to be intrinsic biological and biosocial limits upon the extent to which the chemically induced moment of pleasure can be sustained and woven into a fabric of viable social life. We can readily give up our lease on reality, but apparently it is difficult to gain ownership of paradise!

The use of drugs thus provides a striking metaphor to think of man and his purposes--although one can surely find better myths and parables for these issues. Rather, it is the fact that drug use tangibly affects and influences both personal and social regulations that evokes these general thoughts and motivates this brief essay focusing on non-pharmacologic factors. My purpose is perspective on these--since the conference is obliged to come to grips with a number of explicit problems. Nevertheless, the agenda of drug conferences inevitably drifts, and I believe that the slippages in focus refer ultimately to these fundamental issues of man's intrinsic organization and capacities for work, pleasure, avoidance, adaptation and invention.

If drugs did not simultaneously affect both private experience and public behavior, if their use did not provoke value judgments about pleasures; if their use did not provoke anxiety about the control of expected behaviours--about whether or not drugged or drug-seeking individuals would in all probablity respond reliably to the social signals and cues by which we influence and regulate each other--the adjudication of drug policy would not be as emotional and complex as it is. Further, if we were not as entranced by the power that drugs can endow to the individual who gives or consumes them--the power to obliterate pain or ignore social consequences, to change the world in a single gulp and thus privately revise its meanings and demands--if these features of human nature and social life were not evoked by drug consumption, we certainly would be less concerned, perplexed and panicked by the events of the moment.

But the point is--and it is regrettable to reiterate such banal profundities--that drug use, as well as misuse and dependence, impinge on belief systems. Our topic deals with drugs, but also inevitably with persons and social groups and their processes and implementing institutions. These include the gamut of economic regulations and transactions, legal responses and devices for recognizing and regulating the entire network of drug invention, manufacture, distribution and licit, formal and informal consumption. This complex linkage of drugs with individuals and social regulations can best be approached, I believe, by agreeing that there are no originally evil people, and by recognizing that there is no such thing as a drug which represents either pharmacological original sin or pharmacological original perfection.

We deal, rather, with neurobiology, psychology and social

psychology. These dimensions operate at every level of the analysis of drugs--from the molecular and membrane events to the discernment of affected physiological systems and subjective effects, to an analysis of the various uses made of drug experience. At each of these levels the psychosocial factors may intrude. For example, the turnover of brain enzymes and substrates induced by environmental stress can change parameters of drug action. Put simply, we require ordinarily toxic doses of sedatives in states of catatonic excitement or panic; the brain's response to environmental input and signals thereby changes an aspect of drug action. But social and psychological factors intrude at every relevant point of interest: from the very definition of what chemicals are considered fit for consumption; to who should consume and provide drugs; to the occasions appropriate for this and the tasks to be performed <u>in</u> a drugged state and following it; to the definition of harm or good, of tolerable and intolerable uses and degrees of acceptable dependencies.

The way the consequences of drug intake are anticipated is crucial. American Indian sages note that their legendary vulnerability to firewater is believed by the Indian--who is, in a way, "drunk" prior to starting his weekend binge. He fulfills the white man's prophecy, as do many addicts and children of whom the worst is expected, an expectation which in a certain sense is shared. For the Indian, to drink at all is to have reached the end of one's rope. It is the time to "let go" and the consequences of alcohol are frequently violence and deep inebriation, rather than courtly pleasantries. While there may well be genetic differences in the response to alcohol and capacity to tolerate it, these psychosocial expectations are crucial. Similarly, the consequences of intoxication are socially defined; whether one is--after a noisy drunken spree--subsequently accepted or ostracized is critical. Whether the event is disapproved but tolerated and accepted as a part of what occurs in life, or whether those who succumb and show loss of control are considered pariahs, helps to determine the meaning of drinking and the uses made of drink in a society. And what to do about tissue dependence is also socially defined: whether one takes more alcohol for a hangover or resorts to stoic endurance, for example. Finally, it should be clear--within obvious dosage limits--that it is the way we hope to manage the effects of drugs, the personal and social uses to which we put drugs, that defines and shapes the pattern of drug affects, whether this is called drug use or misuse.

The physician's patient who wants drugs is attempting to regulate his feelings, aches, tensions--that odd psychosomatic mixture expressing dys-ease--in order to enhance his personal purposes and to get his "job" done. He may wish to have aid in enduring his pain while living and coping with his life's tasks or a stressful situation. He may wish to shift his attention from the constraints of everyday life or from neurotic inhibitions.

Drug regulation is thus closely linked to self regulation--to the regulation of the self and its purposes. If we diagnose disorganizing anxiety and treat with pharmacotherapy, we are helping the patient to manage the effects of a drug for a specific stiuationally and time-limited purpose. In this sense, medical management of drugs is a social system that interprets, defines and reinforces the purposes of drug consumption. When we give morphine for analgesia, this is an agreement--tacit or explicit--between doctor and patient that a drug is being used for pain; we agree not to exploit the euphoric effects of a narcotic analgesic, even though they are present. In general, the patient agrees not to notice these effects and to localize the entire experience to the medical occasion for use.

Informal social systems similarly perform these functions. During a period of intoxication when judgment is diminished, the group's definition defines the way events are to be interpreted and managed during the drugged state. The bridge to reality in a state of intoxication lies in the customs of the group, as internalized by the individual and as articulated by the social occasion. During the peyote ceremony, the Indian family--parents, grandparents, children and neighbors, in the presence of "guides"(called road chiefs)--consume enormous amounts of their sacramental drug. They are told to ignore the perceptual effects, to let their mind be caught up in the drum beat and the droning of the prayers, to yield the activity of deciding what to attend to--the task of coping with detail--and rather to be "in the mind of God." The group structure and purpose thus define the experience and how it is to be viewed. Bad trips are interpreted and treated with group prayers and instructions in order to capture and divert attention away from evil and disturbing thoughts. A ceremony ends the performance which is thus given a special place in life, circumscribed and defined by ritual.

The intrinsic links of belief, drugs and social regulations--

the roots of which we cannot thoroughly analyze here--not only guarantee a regulated journey through a period of anxiety or a period of diminished capacity for self-regulation, but are manifest in the use of drugs to enhance communion and belief. In recreational drug use, we shift our attentions from the day's constraints to find a different order of events, a communion and community. An old religious question is: "Why on this night do we drink this wine?" The father answers, citing tradition; the wine is used to dispose one to regard the occasion in special ways--to put one in a different frame of mind and to share an unstated but understood communion. Similarly with subgroups and cliques, the drugs become symbolically meaningful in marking a special occasion or purpose, utilizing both the fact of an unstated subjective experience which the drugs induce and the very taking of them to create a sense of distinctiveness.

This tutoring function of what to accept and to reject, and to expect from drugs, the sharing and shifting of the function of decision-making, of internal struggles with perceptions, feelings and ambitions, is performed in therapeutic groups and cults of all sorts--including youth cults--and by physicians and parents. The mother's food is given with the attitude of "it is good for one," for pleasure, relief of hunger, or growth and strength. The child cannot see what happens after consumption of food--rather, he "feels" and has this state interpreted for him.

When we cannot explicitly monitor and track events--attitudes and customs thus regulate our response. Our basic attitudes about what we take into our body are thus intrinsically nonrational. This is one root of the emotional pitch which drug issues characteristically engender. Indeed, this fact makes rational drug management somewhat astonishing insofar as we can accomplish it in scientific medicine. If we look at the extent to which rumor and fad govern the physician's behavior with medicine, the extent to which reports of toxicological effects are greeted with alarm before the rational basis for and validity of such reports are investigated, we can appreciate the emotional power that lies behind our approach to pharmaceuticals.

For, from our very origins we are deeply ambivalent. Drugs and chemicals are either chicken soup--and help us to transcend a variety of ails and ills, bringing potency and power; or they are poison--shrinking our potency, threatening our future or fracturing our genetic apparatus, and so forth. Fads for or phobias against chemicals and foods are the visible consequence of this inevitable heritage. It is of little wonder then that when the drug or potion is accepted in a society or subgroup, that we deny many of our fears and, in fact, often trivialize drug-taking to the point that it becomes--at least in medicine--pill dispensing. We seek the drug for our purposes, and if we can ignore the complexities and the weighing of risks and gains, we banalize and trivialize the drug in order to go about getting what it is we want in life.

The fact that we ourselves can manipulate our own intentions and seek retreat or comfort through the aid of chemicals--in brief, enhance our purposes with them--causes every society concern. For, even without drugs, we can dream while we are awake. And this is related to the power of drug dependence and to the intransigence of it; our drug use is secretly locked into our most private, persevering, primitive and peremptory wishes. We find novelty manipulating our own minds and bodies; children whirl and swing and delight in changing their feeling state. But one's feelings are essentially private, even though we communicate *about* them. And if one has learned to gain some advantage from manipulating his private feelings, he will violently oppose or disbelieve attempts to change him without appropriate inducement. The power of denial of reality is enormous and is reinforced by the gain of achieving immediate and private relief. Thus many dependencies are enormously costly; the individual is truly out of control of his habit, but he does gain something. The fact that he has some source of comfort which is somehow within his power to achieve can outweigh a variety of reality consequences, which will not, in fact, be correctly perceived, even though they are acknowledged. It has been noted that when working with obese people, they do not perceive satiety and require help and training to do so--they may be miserable with their fat, but they believe in their own experience of pleasure or relief and need aid in sorting out the imperceptions which impede their contrary attempts to stop addiction.

The power of belief thus is enormous both in sustaining addictions and in overcoming them--we need only watch some cures of alcoholism with religion--which seems as potent as any drug. Thus belief can sweep aside detail and habit and arrange all events under a guiding attitude which is protective. One can then view all the events in his life as designed or for an ultimate good. The power of ritual, on the other hand, is to capture attention and to divert it from brooding uncertainty; thus, one can either ignore detail or focus upon it in order to treat anxiety. The power of self-administration of drugs to manage uncertainty is quite similar; this power derives from the principle of familiarity or constancy. For example, the cigarette is a familiar experience to the cigarette addict; he is, by using his familiar cigarette, exerting control over what immediately to expect; a thin veil of comfort lies between him and external stimuli, a barrier against intrusion and the unexpected. He has the private assurance of a familiar experience in response to the nicotine and the ritual of the cigarette. These barriers and comforts and reassurances are regulatory functions which for the infant are performed by the mother, and all men have an enduring need for them in one form or another.

But neither psychoanalysis nor conditioning theory have really explained why habits once entrenched are so hard to yield, and why they are rarely forgotten. The practical fact is that the management of one's feelings, the treatment of one's ills and ails through self management can be socialized both in healthy development or in special treatment situations with or without the use of drugs.

The medical profession must count on the fact that folk customs and social functions, including religion and recreational drugs, help people to solve some of their dilemmas. The need to diagnose one's own condition, to learn to tolerate pain and interpret it, to find some relief through extra medical resources and at a reasonable point to seek special medical help, is useful. We could not, in fact, as a profession, adjudicate every anxiety to which people are prone. Intelligent self-medication, then, has a social role and tutoring one's clientele in it is a function of the medical profession. Applied to drug abuse, it is clear that some patients addicted to alcohol and other drugs are treating phobias or enormous anxieties, and perhaps are using drugs which are not the drugs of choice. They may, on the other hand--at least some--be able to manage a better economy of self-regulation than the profession can provide. It is necessary, therefore, to recognize that there are both useful and damaging patterns of self-medication.

When the physician is asked to use drugs to enhance athletic performance, to diminish tension so that overtaxing work can be performed, to use anabolic hormones to enhance the body mass of football players, or thyroid--not for glandular disease, but for heightened energy--he is on the borderline of the very behaviour that drugs are frequently used for in the dubious practices of self-administration. The use of "the pill" is a good example of the interface of pharmaceuticals with social values and issues. Similarly, it is necessary to distinguish between using drugs to improve severe disorganizations of the control over attention (as in the psychoses), or in various states of extreme stress, to their misuse--to depriving individuals of the chance to gain strength and experience in managing stress without such aids. These are all sensitive social and ethical problems and involve sensible judgements which avoid the fanaticism either of abstainers or of users.

Nevertheless, medical management of drug use--however these shades of grey are adjudicated--probably is more relaible and is less destructive than self medication. When the heroin addict agrees to methadone therapy, he can be giving up or shifting his power to use drug to manipulate his own feelings. He may shift his attention from hustling for a drug in the streets to more socialized interests. This can be done if the approach is not simply "pill dispensing." Most critically, he shifts the definition of drug-taking from the street definition to that of "treatment" (again, assuming that treatment is really provided); and this psychosocial fact may be a powerful rehabilitative force to the extent that treatment is useful. A society that can meaningfully, reliably and constantly rather than episodically facilitate re-entry and re-socialization, provides a key factor in therapy. Given the intransigence of many confirmed addicts, and their dependency and the problems they create, this is the most difficult attitude to sustain.

But such sustained focus means that the belief systems--and practices--that served us in the past to isolate drug problems, to prejudge people by their drug of choice--these attitudes must shift. Police must know why they are arresting if they arrest, and all the various agencies in government must understand their mission rather than instinctively react with prejudice and moral judgement. It is not that prejudice is not convenient--rather, it is that we have lost that convenience and must make judgements about whether or not we take a whole variety of drugs, of whether or not we wish to use them and how, and what we wish to do about those who are having problems with them or those who are exposed

This tax upon our society is on our own energies and it is hardly to be welcomed; but it cannot be ignored. Because of the link of drugs to belief, it is obvious that governments will be concerned about the outcome of any definition of appropriate use of drugs and wish to influence such decisions, and so too will every value-setting group.

What so confounds the current scene, then, is a breakdown of social definitions of the occasion, sponsorship and use of drugs. The automatic "no's" have diminished, and the convenience of custom as a decision-making device on what to do about drugs or what not to do has disappeared with the mobility of persons and styles and the role of the media in tutoring our tastes. There is an abundance of interesting molecules today whose effects can be used either just to feel different, or--if one is a connoisseur--to feel specifically different. But motives capture the attention of some who experiment, others who engage in various episodes of drug use, or a minority who are in dedicated pursuit of drugs.

So, drugs are a matter of choice, and youth throughout the world have another option among many of the products of technology and nature for risk-taking. Those of us who have witnessed the emergence of drug excitement and interest in the United States in the past ten years--led by the privileged and their media--are aware not only of interesting epidemiologic patterns, but of the self-limited nature of some epidemics of interest and use. We are also aware of the necessity to reorganize how it is that parents and children, professions and children, governments and the health professions (the latter having lagged) can communicate to articulate and regulate drug attitudes. This is where the issues lie in coming to a consensus about the utility of drugs in our lives. We have had to modify the tendency to blame or deny and isolate. Our task has been to help to mediate a transition in which custom and authentic expertise can interact to again adjudicate and competently guide the regulation of one's experience and attention and one's ambitions and anxieties.

This is a broad educational venture ranging from explicit issues about specific drugs, to more complex issues of values. There are a range of specific constituencies--different audiences--which are involved with a range of different problems. "The Drug Problem" is, in fact, many problems. But by engaging the issues, by testing and trying, we are educating ourselves and a broad range of professionals and consumers. The answers are not at hand. Our methods are imperfect, but as professionals we are beginning to be better able to specify the dimensions of drug use and misuse, and for any particular question posed by society, we are on the verge of being able to outline the inquiry that would be relevant. And for this, both the perspective of history and an appreciation of the complexity of non-pharmacologic factors are crucial.

PROFESSOR JACOB: Thank you very much Dr. Freedman. We have heard an interesting presentation of the non-pharmacological factors of drug dependence. Our next speaker is Dr. Harris Isbell, who will comment on the pharmacological factors.

PHARMACOLOGICAL FACTORS IN DRUG DEPENDENCE

Harris Isbell, M.D.

Professor of Medicine and Pharmacology

University of Kentucky Medical Center, Lexington

I would like to state my bias as being that of internal medicine, pharmacology and 30 years of association with opiate type dependence in the United States. I will now attempt the impossible by trying to discuss the pharmacological factors in drug dependence in ten minutes. This is impossible because there is such a large number of different substances with markedly varying chemical structures, and markedly varying pharmacological actions that are used non-medically and frequently to a degree causing dependence. Some of the most important drugs are listed in the following Table in which substances causing specific types of dependence are grouped together regardless of chemical configuration. Because of the diversity and number of kinds of drugs of dependence, the WHO Expert Committee on Addiction Producing Drugs (1964) defined drug dependence as "a state arising from repeated administration of a drug on a periodic or continuous basis." "The characteristics of such a state will vary with the agent involved and the characteristics must be made clear by designating the particular type of drug dependence in each specific case; for example, drug dependence of the morphine type, of the barbiturate type, of the amphetamine type, etc." (Eddy, et al. 1965)

Table

CLASSIFICATION OF DEPENDENCE PRODUCING DRUGS

I. With Psychic and Physical Dependence:
   A. Opiate Type: Autonomic storm and central nervous system irritability on withdrawal:
      1. Opium and Derivatives: Morphine; Codeine; Heroin, etc.
      2. Synthetics with opiate-like effects: Meperidine (Pethidine); Phenazocine; Levodromoran; Methadone

   B. Alcohol - Barbiturate Type: Convulsions and Deliriums on Withdrawal:
      Ethanol; Barbiturates; Chloral; Paraldehyde; Meprobamate; Glutethemide; Chlordiazepoxide; Ethclorvynol

II. With Psychic and Mild or Questionable Physical Dependence:
   A. Opiate Antagonist Type: Simimlar to opiate type but some differences: Nalorphine; Levallorphan; Cyclazocine

   B. Amphetamine Type: Sleep, Hunger, Apathy, Depression on withdrawal:
      Amphetamine; Dexamphetamine; Methamphetamine; Phenmetrazine; Methylphenidate; Pipradol

III. Psychic Dependence Only:
      Cocaine; Hallucinogens of the LSD Type (LSD-25 and Congeners); Psilocybin; Mescaline; Dimethyltryptamine; Volatile Solvents ("Glue"); Cannabis (Marihuana, Hashhish, Tetrahydrocannabinols; Tobacco; Coffee and Tea.

The universal characteristic common to all types of drug dependence is psychic dependence, a psychological compulsion to take a drug, periodically or continuously to produce pleasure or to avoid discomfort. Psychic dependence is difficult to define and measure but is recognized clinically by alterations in behavior such as undergoing great risks to obtain the drug, obsession with maintaining a supply, use of the drug as the chief means of adapting to life and a strong tendency to relapse after treatment. The secondary and non-essential characteristics (though important when they exist) are tolerance, a diminishing effect on repeatedly taking the same dose of a drug and physical dependence, an altered but reversible adaptive physiological state bought about by repeated administration of a drug which requires continued ingestion of the drug to prevent the appearance of a characteristic self-limited illness (abstinence syndrome).

GENERAL PRINCIPLES

All drugs of dependence are psychoactive and affect the central nervous system, particularly the higher centers, and cause a variety of alterations in subjective feeling states such as elation, a sense of increased mental efficiency (frequency false), diminished awareness of external sensory stimuli, sedation, and distorted sensory perception. Not all drugs that affect the central nervous system cause dependence (the "major tranquillizers," chlorpromazine and reserpine are examples). The necessary psychic effect seems to be a subjective sense of mental stimulation, at least initially. Even drugs used primarily for inducing sleep or anesthesia cause an initial excitation (for example, alcohol, barbiturates, morphine). The physiological and biochemical mechanisms underlying these actions are too diverse and complicated to describe in a short space.

The physiological effects of drugs of dependence are, of course, markedly affected by the personality of the user, his expectations and the environment.

The actions of these drugs and the likelihood of producing dependence vary with the dose, the frequency of administration, the length of time over which the drug and the route of administration. There is a quantitative spectrum in all types of drug dependence ranging from infrequent occasional use (which may not produce dependence at all) to constant use of several doses daily for long periods. In general, the degree of dependence increases with the amount and frequency of use as does the degree of physical harm when such harm occurs. Drugs that cause rapid and spectacular change in subjective feeling states are more likely to cause dependence than drugs that act more slowly (for example, pentobarbital as contrasted with phenobarbital). The rate of change and the chance of causing dependence is partly a function of route of administration and varies in order of increasing rate of change as follows: oral, subcutaneous, inhalational, and intravenous routes. Injection of drugs, particularly intravenously, causes the greatest physical damage because of increased toxic effects and infections.

Tolerance may be due to an increased rate of biotransformation of drug to less active and more readily excretable substances ("biochemical tolerance"), to adaptive changes in cells of the central nervous system ("tissue tolerance") or to both.

Physical dependence is usually attributed to the same adaptive changes responsible for "tissue tolerance." These changes oppose the actions of the drugs and continue unchecked when the drug is dissipated, thus leading to symptoms of abstinence. At one time physical dependence was regarded as necessary criterion for true addiction but it is now known that some of the most serious forms of drug dependence (the cocaine type of example) may not be associated with any physical dependence.

Psychic dependence on a drug is characterized by "a feeling of satisfaction and a psychic drive that require periodic or continuous administration of a drug to produce pleasure or to avoid discomfort. Indeed, this mental state is the most powerful of all the factors involved in chronic intoxication with" dependence producing drugs. (Eddy, et al, 1965).

When physical dependence exists it tends to result in continuous rather than intermittent drug use but physical dependence has greater psychological than physiological importance. The relief of abstinence symptoms by the drug is a potent psychological reinforcer ("God's own medicine") and conditioning of abstinence may occur so that symptoms may recur if the proper stimuli (such as seeing another user) are present and such conditioning may be a potent factor in relapse (Wikler, 1961).

Most drug dependents take more than one drug. In fact, concomitant use of several drugs, frequently with opposite effects is the rule leading to mixed types of intoxication and dependence.

Medical and social concern about drug dependence may be attributable to direct toxicity (acute or chronic changes in tissues, death), psychotoxocity, (drunkenness, irresponsible behaviour) or social pathology (neglect of family reponsibilities, failure to work, obsession with obtaining the drug, etc.).

TYPES OF DEPENDENCE

In the Table the drugs are divided into those causing strong physical as well as strong psychic dependence (Type 1, A. Opiate Type and B. Alcohol-Barbiturate Type), mild or questionable physical dependence (Type II, A. Opiate Antagonist Type and B. Amphetamine Type), and Type III, psychic dependence only. The classification has practical medical importance since withdrawal treatment is desirable for drugs in Group I, is seldom necessary for Group II drugs and not required at all for drugs in Group III. In addition, drugs in subgroup IA are interchangeable within the subgroup (cross-tolerance and dependence) and drugs in subgroup IB are interchangeable within the subgroup. Subgroup IA drugs, however, cannot be substituted for subgroup IB drugs and vice versa.

Opiate Type: Dependence of the opiate-type is the classic prototype of "addiction" and is characterized by very strong psychic dependence strong, (but not fatal) physical dependence and enormous tolerance. The opiates and their synthetic equivalents produce elation, pleasurable relaxation, drowsiness, sleep, and dissociation of the psychic reaction to pain from its pain. Drunkenness does not occur. The opiates depress responses mediated through multineuronal arcs, stimulate responses mediated through two-neurone arcs, and stimulate supraspinal inhibition of peripheral sensory transmission. Hyperactivity in the multineuronal arcs initially depressed by opiate develops and opposes an action of the opiates (accounting for tissue tolerance). Persistence of the hyperactivity after withdrawal of drugs accounts for symptoms of abstinence (Wikler, 1948). These changes may be associated in complex ways with synthesis and release of neurohumors within the central nervous system but our understanding of these phenomena is rudimentary. Parallel paths of nervous transmission that are not depressed by morphine may be activated and "hypertrophy" accounting for tolerance and physical dependence (Martin, 1970). Biochemical tolerance is of little importance in opiate-type dependence.

Withdrawal from physical dependence of the opiate-type is characterized by an autonomic storm (sweating, pipillary dilatation, fever, etc.) and central nervous system irritability (twitches, muscle aches, insomnia). It is self-limited and not dangerous to life. Abstinence phenomena from opiates are precipitated in acute and potentially dangerous form by the narcotic antagonists nalorphine, naloxone, etc. Treatment of physical dependence of the opiate-type is a simple procedure involving merely reduction of the drug of dependence or a substitute with methadone being most suitable (Isbell, 1971). The high degree of tolerance developed to opiates and the high degree of cross-tolerance of one member of the subgroup to all others is the basis for the methadone maintenance method of management of opiate dependence.

Alcohol-Barbiturate Type: is characterized by strong psychic and physical dependence and by moderate, but limited tolerance. These drugs all cause dose-dependent elation, euphoria, diminished controls of behaviour, motor incoordination (drunkenness); sedation, sleep and coma. Ethyl alcohol has the most deleterious effects to body tissues of any member of the group. Tolerance is both biochemical and adaptive. These drugs all increase the electrical threshold for activation of all kinds of nerve cells at all levels of the central nervous system with the cells of the cerebral cortex (concerned with thought processes) and the reticular activating system (concerned with consciousness) being the most susceptible. Signs of abstinence include nervousness, insomnia, convulsions and delirium. Physical dependence on barbiturates is associated with prolongation of after discharge in neurones and has been hypothesized to be due to "denervation hypersensitivity" (Jaffe and Sharpless 1968). Unlike abstinence from opiates, abstinence of the alcohol-barbiturates is dangerous to life and must always be treated by gradual withdrawal of the drug of dependence or an equivalent drug.

Opiate Antagonist Type: is not of great importance since there is no large illicit source of these drugs. It is characterized by moderate to low psychic dependence, moderate tolerance and low-grade physical dependence somewhat similar to physical dependence of the opiate type.

Amphetamine Type: is characterized by strong psychic dependence, marked tolerance and a questionable kind of physical dependence. The most damaging form physically results from intravenous injection which produced great elation, an orgaistic sensation termed a "rush", a feeling of great mental and physical efficiency, insomnia, nervousness and eventually a reversible paranoid psychosis. Intravenous injection is associated with sudden death and many infectious complications. Withdrawal of amphetamines is followed by sleep with increased rapid-eye movement (REM) electro-encephalograms, hunger, apathy and psychic depression. These symptoms have not yet been shown to be due to a specific disturbance in brain function and may be due only to lack of sleep and food.

Amphetamines presumably act by displacing biogenic amines (norpinephrine) from storage sites within nerve cells.

Cocaine Type: is characterized by very strong psychic dependence, no tolerance and no physical dependence. The clinical features are very similar to those of the amphetamine and, like amphetamines, cocaine may act by increasing the level of free catecholamines in the central nervous system.

Hallucinogen (LSD) Type: The hallucinogens (psychotomimetics, psychodysleptiques, psychedelics) cause dose-related sympathomimetic stimulation (dilated pupils, increasing sweating, fever, goose flesh) and sensory perceptual distortion ranging from greater awareness of sensory stimuli to hallucinations. Dependence of this type is characterized by varying psychic dependence, marked tolerance (probably but not certainly due to tissue tolerance) and no physical dependence. Drugs frequently are used in hope of having a transcendental, mystic experience. Dangers include temporary psychoses with suicide occasionally occurring and precipitation of a chronic psychosis. Drugs are usually taken orally and not continuously because of the development of tolerance. The psychic effects seem to be best explained by impairment of central nervous system mechanisms that enable the person to suppress unimportant sensory stimuli. The individual is then flooded with a flood of sensations that he can no longer ignore.

Cannabis Type: is characterized by low to high psychic dependence, no great degree of tolerance in man (requires reinvestigation) and no physical dependence. Materials used range from cannabis leaf (marihuana)to the mechanically more concentrated forms of ganja and charas. The most active material is cannabis in the 1 - Δ^9 - trans-tetrahydro-cannabinol (Δ^9-THC) and the intensity of effects are related to the concentration of Δ^9-THC in the material used, which varies markedly from 0.1 to 10 percent by weight. The drugs are nearly always smoked. Cannabis and Δ^9-THC cause increased pulse rate, elation, euphoria, increased sensory awareness (in sufficient dose increasing to hallucinations), sedation and sleep. The mechanisms underlying these actions are still not well understood. Dangers include transient psychoses. The effects of heavy consumption are unknown.

Tobacco, coffee and tea: Though these are drugs of dependence they do not cause significant behavioural toxicity and are generally accepted in most cultures.

REFERENCES

Eddy, N. B., Halbach, H., Isbell, H., and Seevers, M.H. (1965) Bull. Wld. Hlth Org. 32, 721 (Drug Dependence: Its Significance and Characteristics).

Isbell, H. (1971) In: Beeson, P.B. and McDermott, W., eds. Cecil-Loeb Textbook of Medicine, 13th edition, pp. 131-146 (Drug Dependence Addiction, and Intoxication).

Jaffe, J.H. and Sharpless, S.K. (1968) In: Wikler, A. ed. The Addictive States vol XLVI, Research Pub. of Ass'n for Res in Nerv. and Ment. Dis., pp. 226-246, Williams and Wilkins, New York (chapter IV. Pharmacological denervation hypersensitivity in the central nervous system: a theory of physical dependence).

Martin, W. R. (1970) Fed. Proc. 29: 13 (Pharmacological redundancy as an adaptive mechanism in the central nervous system).

Wikler, A. (1948) Am. J. Psychiat. 105:329 (Recent progress in research on the neurophysiologic basis of morphine addiction).

Wikler, A. (1961) Brit. J. Addict. 57, 73 (On the nature of addiction and habituation).

WHO Expert Committee on Addiction Producing Drugs. Thirteenth Report (1964) Wld Hlth Org. Techn. Rep. Ser 273, 9

PROFESSOR JACOB: Thank you very much Dr. Isbell. We will have now a short discussion period, before proceeding to consider other factors of drug dependence.

DISCUSSION

PROFESSOR KIELHOLTZ: You said nothing about the flashbacks in the hallucinogens. Can you say something about this problem?

DR. ISBELL: Professor Kielholtz asked about the "flashback" phenomenon. By this is meant a recurrence of some of the experiences that one experienced under a hallucinogenic drug, or under cannabis at a point later in time when no drug is present in the body and no additional drug has been taken. It certainly is not an effect which you can contribute to direct pharmacological action of the drug. The drug by then has long ago disappeared so that it must be due to a psychological mechanism and not a pharmacological one. We don't really understand this "flashback" phenomenon which is usually rather easy to manage by firm reassurance.

DR. EVANG: I have two questions to Dr. Isbell. Firstly, you said that the tranquillizers did not produce great problems as far as dependence is concerned if I understood you correctly. I wonder if this is the experience also in other countries; and I would question those psychiatrists who may be here. We seem to have heard reports to the effect that some of these tranquillizers especially the valium-librium group, have unfortunately rather strong dependence producing effects in some patients. (from the floor.... "these are called minor tranquilizers") I see - well, there may be a question of terminology as to what are tranquillizers and what are not. My other question is of more, should we say, complicated nature. You have listed very nicely the drugs according to psychic and physical dependence, and I enjoyed the clear statement in your paper that it is, of course, the psychic dependence which is the important and fundamental. I think this has to be brought home, also to the public. Now, speaking of physical dependence, what criteria do you use for saying that a drug creates physical dependence? Is it the presence of abstinence symptoms which need treatment? Is it the presence of abstinence phenomena that do not need treatment? Is it the presence of tolerance, or is to a combination? And, is this not again misleading the public and the patient? Is it that we, by building on the old morphine model of dependence and withdrawal symptoms, are honoring practically only the somatic symptoms in the question of abstinence phenomena? Everyone who has had to do with a person that has used drugs that were supposed not to produce physical dependence at all, for example Cannabis, has been struck with the fact that he or she certainly presents an abstinence syndrome, sometimes a serious one indeed. The person cannot function until he gets more of his drug or another solution has been found. Psychic dependence produces eo ipso psychologically and socially an abstinence syndrome which is sometimes more important than the somatic. And finally, Dr. Isbell, you did not comment on the "glue"type drugs which you have listed under the "psychic dependence only" in the Roman III in your paper. Chemically these substances are very closely related to alcohol and the type of acute and chronic poisoning caused by them resembles that of alcohol very much. Is there a special reason for not putting them in the same group as alcohol?

DR. ISBELL: Thank you Dr. Evang. I think the question about the tranquillizers is well put and I do believe that it is a question of terminology and perhaps it would be best if we used a term other than "tranquillizers." By "major tranquillizer" I mean drugs of the reserpine, phenathiazine and butyrophenone class that are useful in treating endogenous psychoses. The term "minor tranquillizer" is a pharmaceutical company term. Most of the so-called minor tranquillizers such as "meprobamate", "librium", "doriden" and so on have turned out to be sedative hypnotic drugs similar to barbiturates. I do not call them tranquillizers at all. The major tranquillizers do cause the effects. Opiate addicts always have a very descriptive term. For a drug to be good, it had to have what they called "drive" or "push". I would point out to them that that stuff drove them into bed. And they would say "yes, but felt like we were being more active and more productive, even though we were in bed." A drug has to have that quality of "drive" and euphoria before it will really be misused. But it is the major tranquillizers not the minor. The minor are listed in the group 1A. With respect to psychic dependence, there is of course a psychological abstinence syndrome - a compulsion to get and take the drug is there and drug dependents are uncomfortable without their drug. By physical or physiological dependence we mean something that is associated with signs or physiological events that are measurable in physiological or physical terms, temperature, respiration pulse, the threshold for elicitation of reflexes etc. I tried, I thought very hard to put physical dependence in what I think is its proper place. It is something that is important to the physician, because he may have to treat it and something which has very great psychological importance - far greater than its physiological importance. However, with respect to the association of tolerance and physical dependence, we know of only two kinds of physical dependence - the opiate type and the alcohol-barbiturate type - and both are associated with tolerance. However, we know of drugs to which tremendous grades of tolerance develops, but which are not associated with any physical dependence. LSD is an example. So tolerance and physical dependence do not necessarily go together. Physical dependence is described in terms of abstinence but in terms of things that can be measured objectively. I'm being a little bad here, because I think psychological events can also be measured but not with a chemical balance.

With respect to the question on "glue." This I think pharmacologically would probably have to be put into the alcoholic barbiturate group. But on the whole, inhalation of volatile solvents (glue) is generally a kind of a binge thing and also currently is behavior mostly seen with very young boys and tends to disappear or to be supplanted either by no drug seeking behaviour, or by use of some other drug at a later point in time. Probably because of the nature of the drug, chronic intoxication of this sort would be almost impossible. The main danger on "glue" is sudden death, generally from asphyxiation.

PROFESSOR DENICKER: I should like to ask Dr. Isbell two questions. First, where does he place the dextromoramide type of analgesics? Is it under opiate dependence, under amphetamine dependence or somewhere in between? The second concerns naturally cannabis. We are not yet engaged in the battle of cannabis. It will come - but I should like to know whether the concentrated resin of cannabis has its place in a special category of mild dependence or whether it should be placed with the hallucinogens, for since the 19th Century all the symptoms of hallucinogen intoxication were described in relation to the cannabis resin and not the cannabis leaf.

DR. ISBELL: Dextromoramide is a typical synthetic morphine like drug and would be classified group 1A - the second category of synthetics. I wonder if you were thinking of dextromethorphan, by any chance? Because dextromethorphan the dextrorotating isome of lenomethophan is an anti fussive drug, which in high dose is an hallucinogen. Dextromoramide is an ordinary everyday morphine like drug - one can suppress abstinence with it and one can produce direct addiction and so on.

With respect to this other question - whether cannabis should be classified with hallucinogens or in a special class. I think the pharmacological evidence that we have indicates that cannabis differs from LSD type hallucinogens in that cannabis has strong sedative actions, does not cause pupillary delatation, increased tendon reflexes, etc. In addition there is no cross tolerance between LSD and cannabis. For these reasons we classify cannabis. Cannabis is a very unusual drug in as far as I know you cannot put it in any group - except with itself. Cannabis is cannabis and is not nitrous oxide.

PROFESSOR JACOB: I am afraid we have no time for more discussion and I regret it very much since I had myself listed five or six questions. I nevertheless should like to stress two points. The first one is a further aspect of the interrelation between induction of drug dependence and doses of the drugs. In animal experiments, pharmacologists were able to develop dependence with almost all major drugs producing dependence in man, provided appropriate doses were used. Quantitatively, there were species, strain, and individual differences. Some strains or some individuals were even found to be particularly resistant for some unknown reasons. Extrapolation of these findings to man, and as long as we do not know the biological factors of resistance, would imply first, that most individuals would become dependent to a drug, if sufficient doses are administered and second, that every one is to be considered at risk once he has begun self administration.

The second point is related to the existing distinction between psychic and physical dependence. In fact, methods are being developed to assess psychic dependence in experimental animals and it cannot be excluded that refined techniques will eventually enable us to measure objective parameters corresponding to the psychic drives : when this is done, the distinction will fade.

DR. ISBELL: Yes I would agree that the matter of psychic dependence and its measurement is of very great importance and a great deal of very useful work is going on. I might mention the work of the group at the University of Michigan, in which they find that the drugs that the monkey will spontaneously begin to take when he has an indwelling catheter in his heart and a lever to press for injections, are the drugs on which man become dependent. The drug that monkeys like the most are cocaine and the amphetamines. The monkey will just press, press, press for those particular drugs. Schuster has described various schedules of reinforcement by which it is possible to get some idea of the force of the drive for this drug reinforcement. Such measures can be made to a lesser extent with smaller animals, such as the rat. There is a great deal of work going on in the development of methods for measuring psychological dependence in the lower animals. It is a very complicated area and I think we are going to learn a lot.

PROFESSOR JACOB: We shall now move to consider the social and cultural factors of drug dependence and I call on Mr. Shlomo Shoham.

CULTURAL FACTORS ASSOCIATED WITH CANNABIS USE IN ISRAEL

Shlomo Shoham

Director, Institute of Criminology and Criminal Law

University of Tel Aviv, Israel
**

INTRODUCTION

I am afraid that my remarks are going to be a bit parochial as they are mainly based on research done in our part of the world. However, I do feel that the findings might transcend in relevance to our region for two reasons. One is that we are situated between one of the largest cannabis producing areas in the world and this is to our north - and to our south are the traditional users of cannabis, as well as origin, as being notorious recently as one of the largest exporters of cannabis. Well, of course, I am relating myself to hashish only, which is the commodity in our part of the world which as you know the tetrahydrocanabinol content is something like eight to ten times higher than marihuana.

Our lesson is mainly sociological and anthropological. As you know anthropologists classify cultures that are more meditative and mystical, tend to self effacement and so on and so forth and on the other extreme we have the tool oriented cultures - the doers, the builders - those imbued with the protestent ethic or similar ethics in other religions or other cultures and the idea there is the possibility of the classifying of the various ethnic groups along this continuum. Now one of our observations is that the use of cannabis with the single oriented cultures is not that harmful to the goals of these cultures as it is with the doers, with the tool oriented cultures. We might visualise that the mid-eastern peasant comes home and has his water pipe full of hashish and relaxes to the torch singing of his favourite singer for hours and hours and this would not harm the express goals of his culture. However, in a society or a group or a culture of doers when cannabis is used to some form or in an extensively or consistently this might and to our mind does play havoc to the whole structure and cultural goals of these groups or societies. Another observation in this context, is that with the single oriented groups, there are cultural and traditional factors which contain the use of cannabis and it actually does not go out of hand it remains on a low level. However, with the tool oriented groups, with the doers, who have not previously been exposed to the use of cannabis, the chances are that the excess or the abuse of the use is quite imminent. This is rather similar to other experiences and research on alcohol for instance - on groups that have been traditionally or religiously prohibited against the use of alcohol - once they are exposed the chances of excess are higher.

Another finding is that some years ago the use of cannabis was confined to deviant and esoteric groups, criminals, outcasts, all kinds of retreatists as we call them in our jargon. Today, the trends towards use is more towards the middlerange - the middleclass - and it is spreading in a way that in some groups it might eventually become the model thing to do. It is not there, however; the trend is away from deviants, away from occasional use of deviant groups towards the center groups.

Our evidence is that there is a kind of maturation as we should call it from the use of the cannabis which is a soft drug as it is classified, towards the harder drugs and here again the transfer is more related to a social psychological to a self image, to a socialogical kind of pressure or factor and the willingness of the individual to be exposed to the experimentation with the cannabis, with the soft drug and eventually it is either not enough, or boring or for whatever good pressures he might get out of it - he might go over to the use of harder drugs. Another finding which ties up with what we have heard this morning from Professor Edwards is that the sheer prohibition of the use of cannabis exposes those of the middleclass or the so-called good boys or good youth to the underworld or its equivalence in so far as the smuggling and the provision of the drugs is exclusively in their hands. This in a sense would not only make a certain equivalent with the Prohibition era in the United States but all the theories that we have in criminology, the so-called differential association, the learning by association with the deviant groups is that the acquisition of cannabis might be involved eventually with other types of deviants and in a sense we do have here again the socialogical or the philosophical or if you wish the moral problem but basically the crucial question is the attitude of the public at large, both legal and social towards the cannabis user.

There are no absolutes and there were no absolute kind of prohibitions. We know that any kind of prescription in our criminal law at one time or another was either commendable or legal; we know that patricide was practised among the eskimoes, we know that prostitution was sacred among the Cannanites. We know that incest was practised among the ancient Egyptians and some even say that the incense in the biblical temple was actually a narcotic. In other words, the question of the legality, the question of the attitude of the public at large towards the use of cannabis is in our minds even more crucial than the actual effect of the drug on the individual or the group. In other words, the labelling process, whether to label a certain activity as illegal, deviant, conforming or law-abiding is all important. Finally, we do have also the evidence that the use of cannabis has been or is used by a large segment of the youth in a population not as a thing in itself, that is, not for the narcotic effects of the drug itself, but as a kind of symbol of revolt - the generation gap, the youth unrest and so on and so forth. In other words, the use of cannabis has become more than a narcotic, but rather as a kind of a political, social and educational tool. However, just before I left home, there was a very interesting case which is also interesting in many other ways. One of the youth leaders who is also one of the leading artists in our country, came out with a kind of confession and I'll paraphase just one section of it: "....I, that have been an alcoholic and tried most of the drugs many times - hashish, opium, cocaine, LSD, I am willing to assure you from my personal experience that the use of any one of the narcotics does not increase creativity. On the contrary, creativity is decreased, coordination is poor, sight is unfocused, memory is blunted, sense of time is lost, a second looks like an hour and a minute like a day. Self assurance increases without proportion to actual performance - one becomes pompous and looses self criticism which is essential for creativity. I think that the worse punishment for the artist would be to record him while he is 'stoned' and then let him hear himself when he is sober again."

Well this is a kind of case study to most of our findings, about the physiological, psychological and sociological effects of cannabis. It might even point out, not as a trend, that it might be that eventually we shall reach to a point where the over enthusiasm and the over indulgence with cannabis might go down the slope. In other words we might sociologically experience a kind of a trend that we are climbing up to the mood but then eventually we might go down. This is just one case which I think has its equivalence in the United States and all over the world.

At this time, I should like to illustrate some of the principles which I have just enunciated as they apply to Israel.

PATTERNS OF CANNABIS USAGE IN ISRAEL

Certain patterns of cannabis usage have developed in Israel as a result of Israel's unique cultural, historical, and geographical characteristics. Because this is essentially a cultural phenomenon, this paper will examine cannabis usage in a number of subcultures and place it in cultural and historical context in order to achieve the greatest clarity.

The cultural diversity of Israel's population of 3,000,000 is very great. Of the total population, approximately 2½ million are Jews, 314,000 are Moslem Arabs, 73,500 are Christian Arabs, and 34,600 are Druzes and members of other small cultural groups. By the end of 1969, foreign-born Jews represented 56% of the Jewish population. This percentage has steadily decreased in relation to the percentage of Israel-born Jews since the establishment of the State. Among their numbers, however, are the pioneers who helped to shape the social and ethical character of the emerging State. Many of these pioneers were Jews from Central and Eastern Europe who settled in Palestine during the first several decades of the twentieth century. Today, of the foreign-born Jews, 51.2% are of European or North or South American origin. Jews of Asian and African origin comprise 48.8% of this total. The majority of these Jews settled in Israel following the establishment of the State in 1948.

A distinction is often made between Jews from Western and Eastern countries. In general, Jews from European countries are considered members of the Ashkenazic community; Jews from North Africa and the Asian countries as members of the Sephardic community. Regardless of the ultimate value of this distinction, for the purpose of this paper the terms are suggestive of cultural differences which find expression in attitudes and behaviour vis-à-vis cannabis.

Many of the Aschkenazic Jews who settled early in Palestine helped pioneer the kibbutz movement. Taking these individuals to signal a certain style among the Ashkenazic Jews, it is possible to say (with only one glaring inaccuracy) that they embodied much of the "Protestant Ethic." This is, they were hard-working, future-oriented, secular, and moral. Alcohol played a very minor role in their lives and cannabis essentially none at all. Cannabis usage was not part of their cultural tradition; there was not time for it, and furthermore, cannabis usage was in conflict with basic idiological convictions.

The Sephardic Jews contrast with the Ashkenazic Jews in a number of respects. For example, they were not utopian socialists, their level of formal education was generally lower, they were present-oriented, they came to Israel after the intense period of early pioneering and, significantly, cannabis usage was rooted in tradition in the countries from which they had come. The Arab community in Israel shares with the Sephardic community the fact that cannabis use has been acceptable behaviour - and even legal behaviour- for certain members of their cultures for hundreds of years.

Cannabis, primarily in the form of hashish, can be obtained quite easily and inexpensively in Israel. Israel is in the middle of one of the world's great hashish-producing areas and is a transit country through which considerable quantities of smuggled hashish moves from Lebanon, Syria, and Jordan to Egypt, where there is a large market. The change of borders following the Six Day War in 1967 also helped to make hashish more available by opening the Old City of Jerusalem to Israel residents, increasing contacts between Jews and non-Israeli Arabs, and by giving political control of the Gaza Strip and the Sinai Peninsula to Israel. Israel itself is not a hashish-producing country.

Cannabis is illegal to sell or possess in Israel. Laws regarding use are not unlike those of other Western countries. Offenders (especially first offenders) receive relatively short jail terms, fines, suspended sentences, or some combination of these penalties.

Cannabis in this paper should be understood to mean hashhish; marihuana is essentially unavailable in the Middle East.

In addition to examining cannabis usage among Israel residents, usage among tourists must also be examined; this is necessary because visitors play a role in transmitting ideas about cannabis and in smuggling. Figures in the paper do not reflect patterns of cannabis usage among the approximately half million residents of Judea, Samaria, the Gaza Strip, and the Sinai Peninsula which lie outside the pre-1967 borders.

OVERVIEW

Estimates on cannabis usage in Israel are inevitably inconsistent. The problem of course, is the lack of reliable data on behaviour which is defined as illegal and the latitude for subjectivity which a highly-charged subject like cannabis permits.

In the United States various studies have provided varying but suggestive figures. The Gallup Poll estimated 4% of the population over 21 years has tried cannabis; a Reader's Digest poll conducted by the American Institute for Public Opinion estimated 5% of university students have tried cannabis (Berg 1970); in Canada the Interim Report of the Commission of Inquiry Into the Non-Medical Use of Drugs (le Dain 1970) suggests that eight to twenty million North Americans have at least tried cannabis.

A spokesman for the Israel Police estimates that there are 10,000 people involved with cannabis in Israel. This figure includes dealers in cannabis, although not all dealers use cannabis. A researcher and administrator in the Department of Research of the Ministry of Social Welfare tends to agree with this figure although he feels it may be somewhat higher. This low figure of 10,000 is countered by 100,000 a figure given in a number of newspaper articles and by a District Pharmacist in the Ministry of Health. If the police estimate is accepted it means that there is a very small number of cannabis users in Israel - one third of one percent of the population. The high figure of 100,000 would indicate a rather large number of Israelis use cannabis - 3.3 percent. In any case, this range is indicative of the problem involved in arriving at some kind of reasonable estimate.

Information from the Ministry of Health indicates that the number of cannabis users who have entered psychiatric hospitals has increased. In 1968 out of thirteen people who were diagnosed as drug addicts, one was considered a "hashishnik"; in 1970 the figure was six or seven out of thirty people hospitalized. Although there is an apparent increase in the hospitalization rate, the number of hospitalizations is too small and the psycho-social aspects of diagnosis too complex for these figures to be more than suggestive.

Police statistics as a source of information present some interesting problems. The 1970 Annual Report (Koppel) of the Israel Police states this about drug use:

> There has been a steep and constant rise in dangerous drug offences, the number of case files amounting to 1,678 (1,053 last year), i.e., an increase of 59.4% (one must remember that many drug offenders are not being detected). A steep rise was also likewise recorded in the number of minor and adolescent drug offenders (up to the age of 20) from 361 in 1969 to 657 in 1970. 478 tourists were apprehended this year for drug offences.

Let us look at some of the problems involved with using police figures. In 1970, the number of people arrested for drug offences was 54% higher than in 1969. Was there a corresponding increase in the size of the cannabis-using community? This is hard to be sure of because of at least two intervening factors: 1) the police force is detailing more and better-trained officers to work with drug offences (1969 saw the creation of a drug squad); and 2) police intelligence, particularly in the form of tips from informers, has never been more complete. Many police arrests depend upon this information (especially in the Old City of Jerusalem).

It is interesting that in 1968 the police estimated a cannabis-using population of 10,000; this is the same estimate that is given in 1971. Following police statistics, the cannabis-using community logically should be estimated by the police to number some 20,000.

The Israel mass-media assert that there is an increase in cannabis use in Israel (police statistics may, ultimately be accurate in indicating a trend) and attribute this increase to the influence of visitors and foreign students. In 1967, 69 non-Israelies were accused of drug offences; one non-Israeli was arrested for every twelve Israelis. In 1970, 478 non-Israelis were arrested; the ratio had dropped to one in five. However, the apparent increase is not as great as it seems because there were many more tourists and temporary residents in Israel in 1970 than in 1967.

The police claim that the increase in hashish seized by the police and the army indicates a spread in cannabis use: in 1969, 3, 179 kg. of hashish were seized; in 1970, 4, 303 kg. However, when the amounts of hashish clearly in transit to the Arab countries are subtracted, the corrected figures are 175 kg. and 434 kg. respectively. The figures for Israel-bound hashish may consequently vary according to chance.

Police data must be used with care; it may be more useful in providing demographic and personal data than in indicating the actual use of cannabis. Let us now look at some specific groups.

SUB-CULTURES - YOUTH

Police records show an increase in the number of drug offenders under the age of twenty: 361 were arrested in 1969; 657 in 1970. Juvenile and young offenders constituted 10% of all drug offenders in 1966; in 1969 they constituted 30% of all offenders (Miller 1971).

In a recent study by Peled and Shimerling (1971), 2, 000 Israeli high school students in the four main cities of Israel were questioned about drug use and attitudes. The authors found that 5% of the students claimed to have used cannabis. Only 1% of the students claimed to have used it more than five times. It is revealing that 81% of the responders thought that less than 25% of high school students use cannabis and that only 56% thought the number to be less than 10%. However, when asked how many high school students they knew used drugs, 51% said they couldn't name any. 8% were able to name more than 10.

The majority of students (83%) have negative attitudes toward drug use, according to the study. Their impression is that it tends to disturb proper role functioning. Nonetheless, 20% of the students want to try cannabis and base their motivation on curiosity. 55% said that drug usage is a fashion which will fade. 45% said it is a pattern which will develop. The usual source of the cannabis for a student was through friends; usually the student did not pay for it.

ARABS

The Arabs in Israel as well as their brethren in neighbouring countries have a long tradition of primary cannabis use. The official Israeli attitude is one of respect for the traditions of minority groups in order to avoid inter-group friction. For this reason the police are relatively tolerant of cannabis use among the Arabs.

Aside from their traditional cannabis use, the Arabs are also in the position as the obvious first sources of hashish distribution to Jewish drug dealers.

The proportion of Arabs involved in drug offences has risen sharply since the Six Day War. Much of this increase can of course be attributed to the large number of Arabs that came under Israeli administration as a result of the war. In 1966, 14% of all persons involved in drug offences were Arabs. In 1968, this figure rose to 26%. This is an increase of 267% which is 153% above the basic increase of 114% for all of Israel (Landau 1970, p. 40).

TOURISTS AND FOREIGN STUDENTS

In recent years, particularly during the summer months, Israel has been visited by large numbers of young Western tourists. Many of these young people have used cannabis in their countries of origin. In Israel they are given the opportunity to purchase cannabis at much cheaper prices than abroad. Relatively large numbers of tourists are arrested for cannabis use (478 in 1970). Aside from possible greater cannabis use among tourists than Isralis, there may be other factors involved in their higher arrest rate.

It is very obvious to the police who is a tourist. Young visitors are in particular very visible because of their hair styles and mode of dress. Certain areas, especially in the Old City of

Jerusalem, are known to be hangouts for young tourists who desire to purchase hashish. Tourists also play a large role in distribution of hashish among their friends. If the police are searching for drugs their first impulse may be to stop and search a number of long-haired tourists rather than a random number of inconspicuous Israelis.

In addition to tourists there are also a large number of foreign students studying in Israel at one of the main universities in Tel Aviv or Jerusalem. Gideon Spiegel, the head of the Overseas Students Programme, Tel Aviv University, supplied the following data concerning cannabis use among foreign students.

In 1968 out of approximately 250 students on the programme (all North American), regular cannabis users were estimated as 50%. The same figures hold true for 1969. In 1970 there were 300 students and it was estimated that between 10 to 15% of them regularly used cannabis, although it was estimated that 40% of the arriving students had been regular cannabis users in their countries of origin and that 90% had tried it at least once.

The sharp drop in number of regular cannabis users is attributed to a more stringent selection of students participating in the programme and the announcement of a tougher policy by the university towards cannabis users. Also in 1970, for the first time there were a great number of Israelis living in the dormitories with the oversease students. These Israelis were said to have influenced the foreign students away from using hashish. This would tend to discredit any theories which imply that foreign students play a significant role in spreading hashish use among Israelis.

It was also mentioned that there were many cases of Israeli drug dealers coming to the dormitories to sell hashish. This contrasts with the popular notion of drug transactions always involving foreigners seeking out Israeli dealers.

DELINQUENTS

Before 1967 cannabis use in Israel seemed to be largely limited to members of the underworld. The trend since the Six Day War has been the spread of cannabis use to non-criminal sections of Israel society. This can be illustrated by the following figures.

In 1966, 79% of those persons involved with drugs had previous convictions for non-drug offences. By contrast, in 1968 only 71% of those involved with drugs had previous non-drug convictions (Landau 1970, p. 53). This implies that since 1967 a greater number of hashish users are coming from the non-criminal sections of society.

The Acting Director of the Department of Social Research for The Ministry of Welfare, Yitzak Berman, believes that the delinquent sub-culture in Israel is not a drug sub-culture. One reason for this is that there are no delinquent hashish users who justify their activities on ideological grounds.

Probation officers seem to feel that cannabis use is not widespread. Street workers who are in close contact with both delinquent and non-delinquent youth report that cannabis use is not infrequent in the natural setting.

KIBBUTZ

Since the Six Day War large numbers of foreign volunteers have come to work on the kibbutz. Some of these volunteers have been involved with cannabis. Whether or not they have been spreading its use to kibbutz members is difficult to determine. Kibbutz members, by their very closely-knit social structure, are reluctant to inform on their members to outside law enforcement agencies. This could explain why very few drug arrests of kibbutzniks have been reported.

It is felt, however, that the amount of influence foreign volunteers could have upon kibbutz members is related to possibilities for social contact between the groups. In a kibbutz where knowledge of English is widespread, there would be a better chance that volunteers could influence the attitudes and behavior of members regarding cannabis use.

SEPHARDIC VS. ASHKENAZIC

Sephardic Jews come from lands where cannabis use is primary. Ashkenazic Jews come from cultures where if there was any cannabis use at all, it can only be described as secondary.

In Israel, while Sephardic Jews have always had a higher rate of cannabis use than Ashkenazic Jews, recent figures indicate a contrary trend. The proportion of Sephardic Jews involved with drugs has decreased from 71% in 1966 to 54% in 1968. At the same time, the proportion of Ashkenazic Jews has risen from 9 to 13%. Furthermore, the rate of increase for Sephardic Jews between the years 1966-1968 has been 71% which is 43% below the basic increase for the country as a whole. The rate of increase among Ashkenazic Jews has been 229% which is 115% above the basic increase of 114% (Landau 1970, p. 40).

THE ARMY

Apparently, there is greater cannabis use in the army than is reported. Indeed, one of the justifications for police vigilence in the area of drugs is the fear that its use may spread to the army. Cannabis use in the army is limited mainly to enlisted personnel, although some officers are involved. The army tends to treat cannabis use as a "sickness" and soldiers who are apprehended are sometimes hospitalized.

OTHER GROUPS

Material relating to groups such as artists and intellectuals is highly impressionistic. Cannabis usage seems to vary considerably with an apparently greater usage among the young than the old.

CONCLUSION

One might ask why the recent increase in cannabis use has not reached epidemic proportions in Israel. What subtle processes of cultural change have caused the proportion of Sephardic cannabis-users to decrease relative to their Ashkenazic counterparts? Why is the rate of increase of cannabis use lower for Sephardim than for the country as a whole. Any why do most Israeli youth, both Sephardic and Ashkenazic, regard cannabis use in an extremely negative way? Perhaps some insight into this question could come from the following reflections.

It could be that the Sephardim in their efforts to assimilate into Israel society are rejecting everything that reminds them of their Eastern origins. This may also apply to the use of cannabis.

The Ashkenazim on the other hand, are sure of their place as members of modern Western society. They can afford the luxury of dabbling in exotic Eastern customs without jeopardizing their position as fully assimilated Israelis.

Perhaps the reason why cannabis use among both groups is not as high as might be expected lies in the history of the State of Israel. It may be that the highly idealistic and somewhat Puritanical attitude of the founding fathers are not quite as dead as some people would say they are. Three wars and continuing political tension have probably put a limit on how far the Israelis are willing to deviate from these original norms.

Although the youth of Israel may see themselves as part of the international youth sub-culture - and are affected by its music and movies - the political and social realities in Israel seem to preclude any full-blown realization of the western drug sub-culture.

** This paper was prepared with - Nechemia Geva, Marty Laskin, Avraham Lehrer and Richard Marcuse: Assistants, Institute of Criminology and Criminal Law, University of Tel Aviv, Israel.

PROFESSOR JACOB: Thank you very much Mr. Shoham. The topic is now open for discussion.

DISCUSSION

DR. GELLHORN: I would like to get some clarification for I have heard both from Dr. Edwards and Dr. Freedman of the importance of behaviour as it relates to society, increasing the anxiety of society and leading to the development of various forms of legislation to control the use of drugs. Then Dr. Isbell in his discussion of cannabis indicated that in his estimation the major problem with cannabis was not physical dependence or the psychic dependence that existed, but that the major problem was in the occasional bad "trip". You have stated, Mr. Shoham that there is a maturation from cannabis to hard drugs and I would like to know what the evidence is for that. But I also thought I heard you say that the use of cannabis in many instances was an expression of revolt of youth and by implication, therefore, that a hard legislation against cannabis might very well increase its use. So I wonder if I could get clarification to what seems to be a contradiction.

MR. SHOHAM: Let me take the first point. In what sense can we detect that cannabis has become some kind of symbol? In our observations, we did find out that as the use of cannabis is a group phenomenon mostly, whatever the sociological or psychological reasons for this. Cannabis, for some reason, has been contrasted with the so-called alcohol culture of the older generation, the two or three martini generation or the excess use of alcohol by the parents. Drugs, especially cannabis have therefore been adopted by youth as a kind of symbol of their fight and their conflicts with their elders just because, inter alia, their elders did not use cannabis or used another drug, or they labelled the use of cannabis as bad. This is just a suggestion, a hypothesis. I'm trying to explain or to suggest why this has been said. Among students it has been pointed out that the use of the drug itself is not only for the narcotic contents but as a symbol of their revolt. The question is whether this contradicts or not with the ability to legislate against cannabis. I really do not see any contradiction. The law as seen by the youth unrest movements is the tool of the establishment and obviously is against the use of cannabis. In other words, "you the elders, you the groups that run society, you are legislating against cannabis and thereby making it a crime and forcing us to expose ourselves to deviant groups, to those who deal with the smuggling of cannabis. And because it is illegal, we have to go over to the underworld and be exposed to this kind of deviance." This is just a rationalization. I'm not doing a value judgment. I'm just quoting. Personally, I'm against this attitude, but it is their value judgment. Therefore, the use of cannabis as a symbol is really not contradictory.

DR. FREEDMAN: I really don't find it that difficult to reconcile but it would be very difficult to listen to if I tried to explain why. Concretely, what you see happening in the United States is that we have discovered that there is pleasure to be rightly had with cannabis. This discovery, and the discovery that law and custom looked upon it entirely differently, was exploited in a variety of ways. The mythology of the establishment, and the almost equal mythology of the kids (often supported by members of the establishment who wished they were young again--and don't we all), simply made communication terribly difficult and at a social cost. If you think of the 1960's as the "great stoned age," what have the kids really discovered? They have discovered that they can enjoy cannabis. They also initially discovered something else: the psychedelic mystique. Historically, this came first. It actually led the way and created drug interest; and our "scientific" press-Playboy, Time, Life, the New York Times--mythologizing "youth" gave them a rubric and slogans under which all kinds of behaviour could hide.

These various behaviors and their value and consequence are beginning to sort themselves out now. We see some people who take their grass at night, very much as one might take a cocktail in order to relax. We also see a segment of the population apparently at risk for drug misuse. If we were ever to educate correctly, I would assume that we would educate that all drugs have a risk and a gain. The hardest thing for drug users to believe, or for anyone to believe, is that you might have a problem with a drug. The experimental distance between pleasures and problems, particularly the problem of dependence, is often great.

The fact that dependence to cannabis does occur was brought home to me by two slogans of the marihuana counter-mythology which helped to keep the two groups apart--an instance of social attitudes acting as a disruptive force and enhancing rebellion. The two observations occurred when "operation intercept" (the frenzied attempt by the Bureau of Narcotics to stop the importation of "grass" from Mexico) was in effect. The Bureau was accused of making people turn to LSD, morphine, and heroin, and there was actually an upsurge in the market for heroin. This hardly accorded with the mythology that enlightened youth take grass with delicacy, enjoying the finer things of life. That part of the counter-mythology was exposed with respect to some of the kids for whom drug taking itself was important. The consequences of the other counter-establishment slogan rested upon the psychological process of generalization: "Pot doesn't hurt me--the establishment is wrong on all drugs." Now, if you can prove by your own experience that your parents are wrong, you can use this as a way of intriguing their attention. And so we found that the many problems, the many issues that were brought up as drug issues were in reality life problems and personal problems. The slogan that there are harmless drugs--pot--reflects the new authoritarianism: my own experience is sufficient guide to tell me when custom does not (such as when parents don't know how to teach about a drug). And that carelessness generalized; for, in 1966 when the rate of increasing LSD use began to diminish, "grass" use increased. There was a generalization: they just lied to us about "pot" and so you can't believe them about pills. Therefore the taking of drugs in its own right became, for awhile, a frenzy. And this generalization can lead people into experimenting with heroin.

What has happened is that the groups have begun to sort themselves out. So we have many people for whom the drug and psychedelic mistique is no longer intriguing. The fun of frightening the parents (as a child has been frightened for some eighteen years, or having a secret now which parents didn't have) has begun to dissipate in some sectors. All of these factors--using to provoke and getting tired of drugs--are operating at the same time; and that's why I think the epidemiology of this is so important to sort out. We are seeing different patterns at different times and at the same time; some of them starting in the rural parts of the United States now, while drug interest and novelty of drugs has begun to fade out, even in San Francisco--and that's the bench mark of everything in drugs!

DR. EDWARDS: The introduction of anthropological thinking is valuable but I would like the anthropologist to insist that cultures are not static. We may once have been a tool oriented society in the West, but God has not promised us that it will always be that way. There was a day when in most parts of the world a man was valued very largely for his muscle. We have moved into a psychotechnology where the man who has the highest intelligence is likely to be most valued. But it is entirely possible that within the foreseeable future we are going to hold in greatest esteem quite a different sort of person, a person who is artistic, who is creative, who is contemplative. The central values of society are changing and concomitant with these changes we may see a change in the acceptance of drug use. One needs in this setting to look at the possibility of a slightly different classification of drugs than the one that Dr. Isbell gave us. A dimension which needs to be introduced into our thinking, is the degree of any particular drug's plasticity. Some drugs are much more plastic than others, much more moulded in their action by social control or by the user's personality. It is unlikely that amphetamines will ever be a widely accepted substance in society because they're not sufficiently plastic in their effect for them ever to be tamed. If you load yourself with amphetamines, you're likely to develop fairly disturbed behaviour. The way that you behave when you're exposed to alcohol is on the contrary very much dependent on your personality, and on the immediate group as well as on the wider cultural assumptions. It is interesting to see analogies with certain psychotic states. Schizophrenia is, in this sense, a rather plastic condition: your type of schizophrenia depends on your personality and it depends on the way society is treating you. Acute mania is not so plastic. One of the better features of cannabis may be that it is a very plastic substance. The bad trip too is culturally maleable. Society has therefore a practical question to ask itself when it finds a young man who has an acute drug reaction. Does it punish him? Does it send him down from University or does it discuss with him how more safely to use the drug? One final point. The question of escalation to hard drugs from cannabis. We need to keep our cultural relativism.

PROFESSOR JACOB: I would like to come back to the pharmacological aspects. I am prepared to accept a proposal for a classification in terms of plasticity. However, I would like to point out first, that in the case of cannabis, the plasticity factor is represented, scientifically speaking by the poor chemical definition that we possess, and secondly, I would like a better definition of "plasticity" before we can accept it as a basis for classification.

DR. FREEDMAN: We have, I suppose, some beginning of a definition in animal pharmacology--again going back to Bradley, Keyes and Elkes in their original observation of the exquisite sensitivity of the drugged animal to the environment in terms of the electrophysiological response. Of course, the very plasticity, I think, and what Dr. Isbell suggested (that is, the loss of filters or categories, or constancies, or what ever you want to call them, for structuring input), probably has a great deal to do with "flashbacks." We can use the model of Janet or any of the 19th Century psychopathologists with a "flashback"; there is a flooding of input--a trauma--to be mastered by re-experience. The most interesting loss of constancy was reported to me by a British pharmacologist--one of the first to take it. He used to commute and would read the London Times. Well, after LSD for about nine to ten months (and I don't know why that magic figure), he could not focus on his London Times because of the telephone poles going by on the railroad--that is, what you normally suppress is now difficult to suppress. I found, in working with these drugs over the last fifteen years with people, that it is very hard to predict just what kinds of loss of control occur. I don't think that they are always terribly damaging, but I think they begin to define what we mean by plasticity--I think this could be put to test very simply.

PROFESSOR J. J G. PRICK (Professor Psychiatry and Neurology, Royal Netherlands Academy of Arts and Sciences): I should like to ask a question to Dr. Isbell. I was impressed by Dr. Edward's and Dr. Isbell's papers, but I believe that it is necessary to realise that the problem of cannabis is basically a problem of chemistry, neuro-chemistry, neuro-physiology and of the structural organization of the brain. I would like therefore to ask Dr. Isbell to speak on the chemical pharmacological approaches to the problem, if he could tell us something about the neurophysiological effects of cannabis on the structural function of the brain. We have heard that it is the chemical substances in the brain which comes into contact with cannabis. Microscopical research does not bring this out. I believe that the sub-cortical part of the brain especially the limbic brain comes under the influence of psychotropic substances such as cannabis. The changes in the behavioural patterns of the cannabis abuser is due to to changes in the activity in the brain and therefore the disturbed social behavioural patterns are probably a question of changes in the chemical transmitters and other chemical processes in the brain cells probably related to protein synthesis.

DR. ISBELL: The question is extremely involved. I believe that I can answer it very shortly by saying - how cannabis acts, what makes it act, where does it act and what happens on long continued exposure to strong concentrations, are matters that we really do not know yet. Our ignorance is largely due to the extremely difficult chemistry of cannabis. The chemical structure of its most active ingredient was only eluciated in, about, 1965, and it is only now becoming available for general research. A great deal of work is in progress on all the questions that you mentioned but it is yet too early to attempt any synthesis of what the effects of cannabis might be either in terms of neurophysiological actions or in terms of the effects on the psychological activity of animals. We still have, even though a large part of the chemistry has been solved - a very tough pharmaceutical question. The Δ^9 tetrahydrocanibinol is not a water soluble substance. It is a resin that has to be dissolved in organic solvents or made into an emulsion with the help of emulsifying agents. The answer then is that we don't yet know the answer to your questions and perhaps in the next five years we will know some of them.

PROFESSOR JACOB: I now give the floor to Dr. Mendelson who will speak on the psychosocial factors in alcoholism.

PSYCHO-SOCIAL FACTORS IN ALCOHOLISM

Jack H. Mendelson, M.D.

Professor of Psychiatry, Harvard Medical School

Director, Department of Psychiatry, Boston City Hospital

The museums of alcohology are crowded with many a triptych illustrating the tripartite significance of psychological, sociological, and biological determinants of problem drinking. The picture varies only insofar as to the degree of emphasis given to any of these variables and depending upon the zeitgeist and the special persuasion of the exhibitor certain portions of the picture are prominently displayed while others may be partially hidden or totally obscured. Simplistic reification of the idea that psycho-social factors are important in the genesis and perpetuation of alcohol related illness is a mundane exercise. The task of specialists working within sociological and psychological disciplines is not merely to persuade sceptics of the relevance of their notions but to provide good working hypotheses and techniques to test them.

Those of us who are convinced that socio-cultural factors must be studied to provide an enhanced understanding of the global problem of alcoholism should develop theories and procedures which are consistent with the best contemporary science has to offer. At the present time there appear to be two areas which should be evaluated, or more properly re-evaluated, in psycho-social research. First is the role of the core or nuclear family in shaping attitudes and behaviors relating to drinking practices and problem drinking. Secondly, is the need to critically examine commonalities in group ethics and behaviors which influence high or low risk potential for development of problem drinking by any individual within the group.

The importance of the role of the nuclear family group in drinking as well as many other behaviour problems is common sense perspective. But the subtleties of this domain are often obscured by gross stereotyping procedures which are usually justified for constructing nosological shorthands and to obviate the chaos which inevitably attends individual case data. For example, it is probably true that a high potential risk for development of alcohol related illness occurs in the progeney of individuals who initially manifest alcohol related disorders yet the qualities of the interactional determinants underlying this phenomenon continue to elude us. This has occurred because, in part, we have compromised a microscopic perception in order to obtain, as I have mentioned previously, more wieldly macroscopic generalizations.

I would call your attention to an art work executed by a sculptor from 15th century Pisa and located on the corner of the Dodges Palace adjacent to the Bridge of Sighs in Venice. The artist who produced this work depicting the drunkenness of Noah had an astute knowledge of the intercurrent medical disorders often seen in elderly alcoholic patients. But, in addition, the artist conveyed the psycho-social aspects of Noah's drinking problem as related to his nuclear family group, his three sons. In the Old Testament text Noah was discovered in a state of intoxication by one of his sons who in turn reported his father's "shame" to his brothers. These two sons professed parental respect by covering the old man with a robe to obscure his drunkenness and debauched state. Apparently Noah did not experience a black-out for in subsequent sobriety he recalled that his two "good sons" did not observe or at least acknowledge his debauchery but his "evil son" made no such pretension. The message conveyed in this Old Testament text is not that drunkenness per se is bad but that confrontation or acknowledgement of such in a parental child hierarchical system is unacceptable.

Many contemporary sociologists and social psychologists have reported that the incidence of problem drinking among Jews is quite low. Is it possible that such reported low incidence is an artifact of faulty case finding? It is also possible that the incidence of drinking problems among heterogenous Jewish populations may be quite substantial. Yet if a common cultural denominator for acknowledging drunkenness particularly in authority figures is strongly discouraged the effect would be a tendency toward under-reporting the over-all incidence of the problem. Moreover if social structures permit management of a drinking problem outside of resources which are known to the public domain the magnitude of the problem could be even further obscured.

My purpose in highlighting the possible confounding factors related to case finding and public knowledge of drinking problems is not to negate the possibility that there are real and important cultural, geographical, and ethnic determinants of drinking problems and behaviour. I believe, however, it is necessary to suspect at least one alternative hypothesis concerning some rather global and perhaps naive assumptions that are promulgated in the literature about drinking behaviours of specific societal cultural, religious, or ethnic groups. If we are to understand the crucial factors mitigating for or against drinking problems in individuals or societal groups we must carefully explore all nuances of psycho-social concomitants with the best techniques that are available from modern behavioural and social science research. In our early explorations in this field we must be equally cautious about accepting attractive and potentially true hypotheses as well as prematurely rejecting unattractive false ones.

In conclusion let me say that there are a number of rather strong movements to initiate programmes of education, if you like propaganda about behaviours relating to drinking and drunkenness. It seems to me that any premature attempt to modify drinking behaviour and problem drinking through informational techniques may yield very poor results. I think we do have an opportunity to see if certain informations that we can provide the general public will be efficacious if we assess their efficacy in a well designed manner. But at the present time in looking at it from the point of view of a naive, and someways onlooker, the status of psychosocial research in alcohol related and other drug related disorders is in its very early phases and while I recognize that people who have committed programmes to action and change often find it terribly frustrating to have people who are more concerned with investigative procedures or research procedures are cautioned with a wait and see attitude. I would hope that before any large scale programmes of education or information are initiated our fund of information about the psychosocial concominance about the alcohol and drug related illness will be enhanced.

CLINICAL RESEARCH ON DEPENDENCE-PRODUCING DRUGS

Pierre Deniker

Professor Agrégé - St. Anne Hospital, Paris

It is impossible in a brief statement to consider every aspect of clinical research on drug dependence that ought to be developed. However, because of the changes in the agents which give rise to drug abuse, stress must be laid on the importance of research in clinical pharmacology with specific reference to man. In this paper the author will touch on only a few concrete aspects of such research as they emerge from experience in his own country, which has recently seen new types of drug addiction develop and has had to organize research in a sphere which had hitherto posed no problems.

1. RESEARCH AIMED AT RECORDING THE ADDICTIONS FOUND

The types of addiction now found, particularly among young persons, raise a dual problem - the complexity of the substances used and the psychological features of the persons involved. A survey must therefore deal with both aspects.

Determination of the Drugs Taken and the Doses Used

Unlike what used to happen with classical drug addiction, the exact nature of the drugs and determination of the doses taken are often not known by the patients, who have no interest in them. It is essential therefore to have available teams of specialists and laboratories in order to draw up an approximate balance sheet.

List of the Products Used

Questioning on the basis of a set form and identification from comparison with plates or samples and the recording of the doses absorbed represent the first stage in the examination.

At the same time, a laboratory must be capable of making a chemical analysis and determining the amounts of the substances in the possession of the drug takers. There will be some surprises as a result: for example codeine may be discovered instead of heroin.

The classical methods of identification, such as microscopic recognition of the hairs from hemp leaves in dust taken from clothing, must not be neglected.

Determination of the Drugs in the Patient's Body

Simple tests, sometimes known as "spot tests", make it possible to detect toxic substances rapidly; morphine can be detected in the urine in a few minutes, for example, but these tests are of hardly any quantitative value.

Methods of extraction and identification such as thin-layer chromatography (TLC) supply quantitative information in regard to pure products and every drug addiction unit should be equipped to carry out this technique.

The more complex methods are more laborious. These include gas chromatography (amphetamines) with modifying reactions (cannabis derivatives) or followed by TLC (lysergamide) and studies of metabolites and metabolism. Only large research centres can have these techniques available and the great variety of drugs used leads to them being restricted to the most important toxic substances.

Determination of Mental State

The enquiry as to the case history from the patient and those around him must primarily determine the mental state before drug-taking and the neurotic, psychopathic or psychotic structure of the personality, the last two being the most frequently encountered in practice.

A study must then be made of the effects of the drugs on the mental state and special attention must be paid to syndrome common to various types of drug addiction: negative syndrome of energy deficit, changes of mood and drug psychoses.

Finally, study of the evolution of the conditions makes it possible to distinguish two main groups: psychoses uncovered or aggravated by drug-taking and toxic psychoses proper.

Prospects for Basic Research

Better knowledge of the toxic features and those which affect psychic structure requires research going beyond the boundaries of clinical investigations.

This research is concerned on the one hand with psychobiological and metabolic changes as a substrate for drug dependence.

It will involve, on the other hand, epidemiological studies comparing populations of drug-dependent subjects and populations of those "at risk", based on surveys carried out in particular in school and university circles.

2. RESEARCH IN CLINICAL PHARMACOLOGY

A distinction must be drawn between study of complex intoxications (multiple drug addiction) and those which provide a better knowledge of the abuse of specific drugs. It would therefore be valuable to give international encouragement to work on relatively "pure intoxications".

The Example of Cannabis

In France, the authorities responsible for research have given priority to studies on hemp and its active principle. This is justified by the fact that it can be consumed in many different ways, whereas identification of the tetra-hydro-cannabinols has only recently been achieved and their clinical effects have not been sufficiently studied, which means there is a certain scientific "vacuum" in the face of propaganda representing hemp as a harmless drug.

In our country, taking into account American work, research has been centred on the pharmaceutical chemistry of hemp, its resin and the cannabinols, and on the clinical studies on healthy volunteers with pure or titrated products.

However, these types of research, while simple in principle, encounter difficulties which also faced research workers in the United States in regard to the solvents used, the instability of the active principle and the effect of repeated administration.

- Alcoholics solvents only take up part of the active principle, whereas oily solvents have a higher content of THC (tetra-hydrocannabinols) but produce slower effects. It therefore becomes necessary to distinguish the results obtained with the different types of solvent.

- The instability of THC (particularly the -9-trans variety), which has to be kept under nitrogen at low temperature, and of the resins is another source of error. The plant seems to preserve the active principles well, but it is difficult to measure them in the plant and the smoke from a few cigarettes in any case gives but little information on the human pharmacology of the THC.

- Finally, while it appears reasonable to call upon volunteers for strictly limited and controlled experiments, it is impossible to approach the conditions of chronic intoxication, hence the value of psychiatric studies on cannabis-taking in the countries where it is common, such as were carried out in Morocco.

Hallucinogenic Substances

The effects of acute administration of the main hallucinogens (mescalin, lysergamide and psilocybine) are quite well known and can be compared with the first studies on hashish to produce a collection of signs common to absorption of these substances, if allowance is made for the doses and psychological background.

However, the investigation of new substances such as methoxy-methylamphetamine (STP) or piperidyl benzylates

(Ditran) involve physical and mental dangers which preclude tests on volunteers. It therefore becomes necessary to isolate the cases found in sufficiently well-equipped research centres.

The Amphetamines

While the clinical pharmacology of these substances can be considered as well-known, the same does not apply by any means to the study of the amphetamines psychoses. Such studies can only be carried out on isolated cases or in certain areas where amphetamine-taking is common, such as Japan or Sweden.

Opium Derivatives

Intoxications with opium derivatives are doubtless those with the longest known history, although the mechanisms of addiction have not been elucidated.

A search for antagonistic medicaments (anti-drug substances), or possibly less dangerous substitutes, is still something for the future.

Association of Drugs

Although these are more and more frequently encountered, their toxicity (e.g. lysergamide + amphetamine) seems to preclude their being studied in healthy volunteers, hence the importance of intensive research in any cases that are detected.

The Appearance of New Types of Drug Abuse

The random search carried out by addicts themselves means that new products are continually being involved, particularly medicaments which are not used for their proper purpose[1]. The important thing is to detect these types of drug abuse as early as possible so that they can be studied from the point of view of drug dependence. That is one of the aims of the WHO International Reference Centre on the Adverse Effects of Psychotropic Drugs, of which we are in charge and which is developing a specimen information form to be sent out by the International

1. Among these can be mentioned drugs used to control Parkinsonism and some tranquillizers associated with alcohol.

Reference Centre for Information on Psychotropic Drugs, which has its headquarters in Washington.

3. RESEARCH ON TREATMENT AND REHABILITATION

Such research is specifically a type of clinical research and justifies considerable efforts in the present situation.

Withdrawal and Disintoxication

Although this stage in treatment raises fewer problems than in the classical types of drug abuse, a proper treatment schedule is still to be drawn up to replace the empirical methods used here and there. The administration of barbiturates (in amphetamine poisoning), and of some tranquillizers, carried a risk that one type of intoxication will be replaced by another.

Preference should always be given to neuroleptic drugs and substances which do not produce euphoria. The use of hemineurine (disulphonate ethane of methylchlorethyl-thiazole) has been advocated but has not been the subject of extensive or controlled studies.

Treatment of Psychiatric Syndromes Resulting from Drug-Taking

A special interest attached to the treatment of syndromes that persist in a more or less latent manner after disintoxication.

We have studied the syndromes of energy deficit common to modern forms of drug abuse. Their treatment requires special efforts on which rehabilitation will depend. In these cases a new use has been found for insulin cures (of the Sakel type or with sudorific doses) which make it possible to establish a true medical relationship with the patients.

Syndromes of mood distortion or a more or less atypical depressive type call for attempts at treatment with antidepressive drugs and Lithium.

In drug psychosis syndromes a different treatment must be given to real psychoses aggravated by brug abuse and toxic psychoses proper and a distinction must be drawn between the two groups. Here again therapeutic tactics have been ill-defined and cooperative and comparative studies would be justified.

The treatment of drug dependence depends to some extent on the treatment of the sequelae which have just been mentioned but thought could be given to specific types of treatment (anti-drug medicaments) which must be looked for.

Rehabilitation of the Patients

This is an essential phase in the treatment in which, for the moment, we are reduced to complete empiricism. It requires particularly urgent psychosociological investigations. The type of institutions for medical or non-medical socio-therapy must be determined and it must be decided whether these institutions should be reserved for drug patients or should receive other patients. In the latter case what type of patients should be admitted?

CONCLUSIONS AND RECOMMENDATIONS

The first recommendation that can be made is to establish rapidly teams of specialists in various domains capable of undertaking the research made necessary by the new forms of drug abuse.

Among the clinical researches necessary a distinction can be drawn between -

1) those aimed at obtaining a better knowledge of complex drug abuse and determining its effects from the biological and psychological point of view;

2) research in clinical pharmacology on the main types of toxic substances. Since trials on volunteers are restricted by considerations of medical ethics, research on specific forms of drug dependence deserves special encouragement at the international level;

3) special efforts must be made in regard to therapeutic research on the various aspects of drug dependence.

PERSPECTIVES IN PHARMACOLOGICAL RESEARCH ON DRUG DEPENDENCE

Dr. T. L. Chrusciel

Medical Officer, World Health Organization, Geneva

The pharmacological problems connected with drug abuse and drug dependence constitute but one, albeit an important, part of the great complex of research and public health problems that arise in connection with drug taking for non-medical purposes.

From the public health point of view it is very useful to consider drug dependence as a communicable disorder. In the past, thinking in regard to drug abuse focused either on the substances taken, on the persons taking drugs, or on the social and cultural mores of the population. The public health approach combines these three aspects, the problems of drug abuse and of dependence on drugs being viewed as consequences of the interaction of three factors: the agent, the host and the environment. This takes into account all the elements involved: the drug as agent, the individual user as host, and the social setting as environment.

The availability of a desired substance at a given time and place is, obviously, a precondition of its use. The pharmacologist wishing to contribute to the solution of problems of drug dependence has several theoretical questions to deal with and the clinical pharmacologist has both theoretical and practical problems. They should both be concerned with the following research issues:

1. Identification of dependence-producing drugs
2. Identification of drug dependence and the nature of the agent involved
3. Basic research, including analysis of mechanisms of action
4. Search for new diagnostic and treatment procedures
   4.1 Search for new agonists
   4.2 Search for antagonists of dependence-producing drugs
   4.3 Search for other chemotherapeutic approaches
   4.4 Significance of agonist-antagonist components in the same agent
5. Evaluation of efficacy of existing treatment techniques which employ a pharmacological approach
6. Search for new drugs to replace the dependence-producing ones in medical therapy
7. Evaluation of effects of long-term taking of dependence drugs
8. Search for links between dependence liability and abuse potential
9. Search for means to minimize the abuse potential of dependence-producing drugs
10. Coordination of research

Now I should like to elaborate on these problems:

1. IDENTIFICATION OF DEPENDENCE-PRODUCING DRUGS

What are the properties of dependence-producing substances which make them desired and actively sought for non-medical use?

Most unfortunately there is no single answer to this fundamental question because (1) there are many substances, possessing different pharmacological properties, which have been used for non-medical purposes, and have produced drug dependence and (2) the effects of such agents vary according to the dose adminstered; the physical, and/or the psychic, state of the user; the frequency with which the drug is taken; its route of administration and whether other drugs have been taken, regularly or irregularly in the past or concomitantly with the substance being considered. Often even the "social setting" the environment in which drug-taking occurs, may contribute to the effect. In earlier times the choice of available psychoactive substances was not large; among those few, one usually became the local drug of choice, e.g. alcohol, opium or cannabis preparations. This is no longer so.

In recent years psychopharmacology has been developing rapidly and the number not only of substances available, but also of research publications on them and on their effects, is increasing in an exponential manner. Even with the help of sophisticated computerized bibliographic services, it is very difficult to keep completely up to date. Despite the abundance of information issuing from research laboratories and clinics all over the world, possibly one thing is certain; the pharmacological bases of drug dependence have not yet been fully elucidated and we are discussing a phenomenon which develops in a way about which we know little, and whose mechanisms of action are particularly obscure. Certainly, there are several theories as to the basic processes involved in the mechanism of development and in the process of drug dependence itself, but the authors of these theories (Martin, Colliers, Paton, Wikler) contend only that they have construed working assumptions and hypotheses and their theories cannot be proven until further research data have been accumulated.

However, the abundant research data already available do permit several general remarks on the pharmacological approach to problems of drug dependence.

Substances or mixtures of substances that are liable to produce dependence must possess some influence on the central nervous system. Their affects in man may be to stimulate or to depress the central nervous system or to bring about disturbances in thinking, behaviour, perception, mood and motor functions; they may or may not be accompanied by auditory, visual or other sensory hallucinations. For a substance to be dependence-producing for a particular user, the effects it produces will always contain some elements which he will interpret as desirable to him, which remain in the memory of the user and invite him to repeat the encounter with the substance.

At the very outset of his study, the clinical pharmacologist dealing with dependence-producing drugs will come up against psychological as well as physiological factors connected with drug taking.

Because of the importance of the dosage and drug-taking schedules and because of the development of and of the variety of action of different categories of dependence-producing drugs, close collaboration is required between a pharmacologist and a psychologist in developing objective, reliable psychological and physiological methods for application in clinical research on the identification of dependence-producing drugs. There is a great need for development of such methods. I should like to underline an urgent need for further studies on methodology connected with assessment of psychic dependence.

The degree of satisfaction desired and the level of pleasure expected depend on complex factors of personality, expectations, educational, cultural and social background, age, sex, and, last, but not least, the motivation for the drug taking of the individual involved. This motivation has been described in detail by the WHO Expert Committee on Drug Dependence which met in 1970.

The modern, sophisticated, well-informed user frequently has a considerable, though limited, knowledge of psychopharmacology. He differentiates between the hallucinogenic, stimulant and depressant drugs. He knows or thinks he knows from his friends in the drug subsculture or from his personal experience which drug can be substituted for another when his preferred drug is not available. He may also know how to use several drugs at a time to achieve the desired level of effect. His action in this connection may be disastrous. Multiple drug use is now spreading all over the world, especially among youth.

2. IDENTIFICATION OF DRUG DEPENDENCE AND THE NATURE OF THE AGENT INVOLVED

The classical pharmacologist has little to say in this matter. However, new and better ways to detect and identify drugs in body fluids are obviously needed and, in particular, there is a great need for simplified methods and techniques. Such methods are needed for diagnostic purposes and for therapeutic supervision of users. Simplified techniques would be useful for epidemiological research. The attainment of this aim depends, of course, upon the development of a substantial body of knowledge on the metabolism and excretion of dependence-producing drugs.

Research should continue on the synthesis of labelled

compounds, which are necessary for the analysis of the absorption, fats, metabolism, distribution, storage and excretion of dependence-producing drugs.

Identification of persons who do not use drugs, although they may be closely attached to drug-taking subcultures is very important. Search for genetic, developmental and biochemical factors involved in abstaining from drug-taking may prove extremely valuable.

There is a great need for more facilities, especially for basic research on drug dependence in man. At the moment only a limited amount of human work can be done on drugs other than the opiate-type.

3. BASIC RESEARCH, INCLUDING ANALYSIS OF MECHANISMS OF ACTION

Basic research should concentrate on a very ambitious and difficult goal: to explain the basic mechanisms involved in the development of dependence to drugs in man and to elucidate the processes in the central nervous system at the subcellular level during chronic taking of dependence-producing drugs.

Much greater emphasis should be laid and greater effort is desirable in research of all kinds on dependence of the barbiturate, amphetamine, hallucinogenic and cannabis type since these are the kinds of dependence about which we know the least.

It is expected that continuously developing knowledge on biochemical processes occurring in the central nervous system on the interaction between the receptor on the one hand and agonists and antagonists on the other will help to indicate new approaches to the pharmacological problems of drug dependence.

More research should be done to explain the phenomenon of tolerance. Analysis of mechanisms of delayed reactions to drug-taking should be undertaken. It is well-known, that after use of lysergide, some mushrooms taken for hallucinogenic purposes and some other substances, a number of individuals suffered a delayed reaction, more or less a recurrence of the initial effect experienced several weeks after drug use. Some patients required admission to hospital for psychiatric care. The mechanism of delayed reaction is obscure and research to explain this phenomenon is needed.

More research should be done on processes indirectly relevant to development of drug dependence. For example, the physiological and biochemical bases of learned behaviour which is undoubtedly involved in the development of drug dependence in man are not elucidated. It is tempting to suggest that the discovery of selective blockers of some processes - particularly memory processes - occur under the influence of dependence-producing drugs might be found useful in the treatment of drug dependence. Consequently an attempt to explain the subcellular processes constituting the basis for memory seems an important contribution.

4. SEARCH FOR NEW DIAGNOSTIC AND TREATMENT PROCEDURES

1. Search for New Agonists

The WHO Expert Committee on Drug Dependence noted in 1970 that "medicine has traditionally endeavoured to establish various treatment goals, ranging from full recovery to minimization of the progress of the disorder, alleviation of symptoms, prevention of complications, and the provision of comfort for the afflicted when nothing more is possible. There is thus nothing new in the concept of providing medical aid for an ill person when a fully effective treatment is not available." The Committee explained further: "The concept of maintenance on a drug of dependence, such as heroin, or on a substitute drug of the same type, such as methadone, has been explored in order to determine if some of the pathological effects of drug dependence could be alleviated without necessarily achieving full recovery."

In this connection the pharmacologist is expected to search for new agonists which may be used in maintenance therapy. More research should be done on agonists having a lesser dependence liability.

It may be possible to contemplate a rigid limitation of the therapeutic use of dependence-producing psychotropic drugs, but only after comparably effective drugs having no dependence potential will be available.

Interesting research is being done with agonists which have some antagonistic properties at the same time. Narcotic antagonists are not devoid of dependence liability (Eddy, 1970), but in several other drug categories substances are known that possess at the same time both agonist and antagonist properties and research on these drugs is a really virgin area.

2. Search for Antagonists of Dependence-producing Drugs

Studies on the mechanisms of action of dependence-producing drugs are of basic importance for the development of antagonists to be used in the detoxication and other treatment as well as in diagnostic procedures. There is a great need for such studies to be undertaken. If a suitable antagonist is found it may be useful in treatment and diagnosis and it may also prove a useful tool in further studies on the elucidation of processes of development of drug dependence and help to clarify the mode of action of drugs of the same category. A search for specifically selective antagonists will be very helpful.

Pharmacological antagonism may be considered as any interaction whereby one substance hinders the biological action or fate of another. A drug that selectively suppresses the effects of another either by breaking a link in the chain of its transport through membranes or otherwise within the body; by changing its action, storage or metabolism; or, finally, by promoting its inactivation or excretion, may be regarded as an antagonist. Antagonists to opiates are indeed already applied, by various techniques, in cases of dependence of the morphine type, e.g., naloxone and cyclazocine in treatment and nalorphine in diagnostic procedures. But antagonists to other dependence-producing drugs, especially those in the categories of stimulants and depressants and hallucinogens, are either unknown or not yet explored from this angle. Consideration jointly by chemists and pharmacologists should be given to which classes of the existing categories of drugs and chemical configurations may be suggested for possible trials as to their antagonistic effects. Research on synthesis of derivatives of known antagonists should continue and pharmacological analysis of new compounds should be immediately undertaken. Screening methods for various antagonistic properties should be developed. On-going research on existing antagonists should be fostered and speeded up. A search to find agents more effective than any known today should continue. A joint, coordinated, industry-wide research effort should be mounted on an international scale. A depot type of antagonist, harmless in itself, would be very valuable.

Ideally, model sets of properties characteristic of antagonistic drugs in the various categories of dependence-production should be described, and such sets should be revised when necessary. Indeed, in all phases of research for new antagonists consideration should be given to the known or hypothetical mechanisms of action of agonists. If these were known it would be possible to search for substances influencing directly the development of dependence. It is conceivable that such substances may prove vitally important in prevention and treatment of drug dependence.

3. Search for Other Chemotherapeutic Approaches

Continuous attention should be given to the analysis of the mechanisms of action of dependence-producing drugs. For certain drugs some clinical trials have already shown the applicability of adrenergic receptor blocking agents in the therapy of drug dependent individuals. In particular β-adrenergic blocking agents have been applied in the treatment of amphetamine addicts in Sweden (Gunne, 1970) with promising results. This approach merits further study and clinicians working with addicts should be invited to do research along these lines, taking into account the results of research on the mode of action of dependence-producing drugs and the increasing amounts of different blocking agents which are becoming available.

Agents that block the release of action or that influence the synthesis of biogenic amines should be investigated as a means of preventing the development of drug dependence. Several approaches along these lines have already been undertaken. Research data on the connection between biogenic amines and the action of dependence-producing drugs are increasing exponentially. Most recently Smith (1971) from the NIMH Addiction Research Centre in Lexington has shown that opiates increase synthesis of catecholamines. It is conceivable that trials to reverse this action by inhibiting synthesis can be contemplated in order to antagonize the effect of opiates.

4. Significance of Agonist-Antagonist Components in the Same Agent

Research that began about 40 years ago on the dissociation of analgesic and dependence capacity has led to the discovery of several agents which possess both agonist and antagonist activities. Some of such substances are being shown useful in medical therapy. However, we need still more such agents possibly more effective in the control of pain and with less dependence liability. The occurrence of psychotomimetic effects after increase of dosage of agonists-antagonists constitute a drawback in therapy but it diminishes the abuse potential considerably. Studies on the agonists-antagonists effects seem to be very promising and are required.

5. EVALUATION OF EFFICACY OF EXISTING TREATMENT TECHNIQUES, WHICH EMPLOY A PHARMACOLOGICAL APPROACH

Evaluation of results of treatment and other techniques should be incorporated in all research on the chemotherapy of drug dependence. The desirability of arranging for long-term objective and systematic evaluation of programmes for the prevention and treatment of drug dependence is very widely accepted. Clinical pharmacologists should play a key role both in the planning and execution phases of the assessment of results of chemotherapy. The WHO Expert Committee on Drug Dependence recommended in 1970 that "...support be given to the setting up of evaluation procedures wherever possible."

6. SEARCH FOR NEW DRUGS TO REPLACE THE DEPENDENCE-PRODUCING ONES IN MEDICAL THERAPY

There are at the moment more than a hundred dependence-producing analgesics and psychotropic substances in therapeutic use all over the world. Some of these are used only in a few countries, while others enjoy a worldwide application. The assessment of therapeutic usefulness and detailed comparison of safety and efficacy of these drugs should continue. Search for new drugs having the same therapeutic spectrum of action but devoid of dependence liability is an important research function of the pharmacologist.

7. EVALUATION OF EFFECTS OF LONG-TERM TAKING OF DEPENDENCE-PRODUCING DRUGS

The information available on the long-range effects in man of numerous categories of dependence-producing drugs is inconclusive and incomplete. Research on the effects of long-term drug-taking on endocrine glands and their functions, various vital enzymatic systems and functioning of autonomic nervous system is practically non-existent. Analysis of the effects of chronic drug use in various pathologic conditions, such as chronic infections, blood diseases and endocrine disturbances is badly wanted by practitioners, who are confronted with practical questions connected with drug-taking by such patients. Because of difficulties of collection of comparable groups of patients, long-term, prospective, collaborative studies should be planned well and in detail.

8. SEARCH FOR LINKS BETWEEN DEPENDENCE LIABILITY AND ABUSE POTENTIAL

The question whether the dependence liability and the abuse potential of a dependence-producing drug are parallel or divergent is frequently raised. A pharmacologist disposes of a well-developed methodology which enables him to determine the physical dependence capacity of a drug with a high degree of accuracy. However, as indicated previously there is no such methodology applicable to a drug's psychic dependence capacity. The scoring of psychic dependence capacity based on self-administration methods in animals is not yet conclusive and further research is needed. The abuse potential of a drug can be predicted by taking into account all pharmacological and clinical data, but is its dependence liability always parallel with its abuse potential? As far as narcotic drugs are concerned such parallelism between physical dependence capacity and abuse potential is well established and the clinical data have supplied ample evidence. All narcotic drugs which have scored high in physical dependence capacity have indeed proved to have a high abuse potential. There is less evidence of such parallelism between psychic dependence capacity and abuse potential, and lack of detailed and objective methods for assessing the degree of psychic dependence, as well as fluctuations in the degree of psychic dependence which may occur, making the judgement more complicated than it seems. Now days, numerous non-narcotic drugs are being abused; some of them have a high psychic dependence capacity but much less, if any, physical dependence capacity. The problem may well be worth reconsidering and clinical studies should follow. Oswald et al. (1971) suggested recently that certain mood-influencing drugs may not be drugs of abuse because of some unpleasant initial effects though they can be drugs of dependence.

Methamphetamine and phenmetrazine can serve another example. Methamphetamine is frequently abused in the United States of America therefore its dependence liability and abuse potential are both high there. It is, however, little abused in other countries, which seems to indicate that abuse potential and dependence liability may be separated. Phenmetrazine has a well developed abuse potential in Sweden and some countries but not in others.

9. SEARCH FOR MEANS TO MINIMIZE THE ABUSE POTENTIAL OF DEPENDENCE-PRODUCING DRUGS WHICH ARE USED IN MEDICINE

Although of low priority from the basic research point of view, such studies have been already undertaken and they should continue. The results of such studies may have a considerable practical impact. The methods applied by the industry to minimize the abuse potential of dependence-producing drugs are: 1. making the dependence-producing substance inextractable; 2. making the use of larger doses of dependence-producing drugs difficult or impossible by adding to the preparation certain other substances, which may or may not be biologically active. In this connection, search for sensitizing agents which could increase the unpleasant effects of particular categories of dependence-producing drugs, e.g. disulfiram, is well justified.

10. COORDINATION OF RESEARCH

Research into drug dependence is likely to be a vastly uneconomical undertaking both because of the scarcity of comparable groups and because of the number of variables involved unless it is pursued on a national or international basis and over a lengthy time space.

Some thought might be given to the possibility of coordination of research at the national, and when required, the international level by specially convened multidisciplinary committees.

Such bodies, taking advantage of local expertise and of the knowledge accumulated by international bodies dealing with drug dependence problems may help to establish research priorities in accordance with the latest state of knowledge and with local needs and possibilities. The close collaboration between such national bodies would greatly help to speed up the resolution of urgent research problems, would enable goals to be established and point out directions of research, and by its interdisciplinary character would greatly contribute to the solving of problems which are on the borderline of various scientific disciplines and remedy the shortcomings of a one-discipline approach.

International medical and public health organizations such as the World Health Organization and international societies, such as CIOMS or the International Union of Pharmacology, may be very helpful in coordinating the work of such national committees.

The indication of priorities in research would be a very important task for the national committees and would help the funding authorities, governmental, or other, to finance research appropriately.

Any research programme proposed by a national committee should incorporate (1) a strong scientific leadership with participation of the scientific community in the formulation of short- and long-term plans and research proposals; (2) major emphasis on and support for training of young research workers and physicians, social workers, and other specialists; (3) substantially expanded support for individual exploratory studies of fundamental biological problems involved in drug dependence.

The programme should present a coherent overall plan and should focus on specific research goals. The plan should be formulated after full consultation with the scientific community to consider its value and its impact on other programmes.

The programme should also incorporate appropriate mechanisms by which advances in drug dependence research may be translated into practical application in the prevention, control and treatment of drug dependence in man.

The activities of these committees should not, however,

monopolize research and should not be an obstacle to sound research which a committee has been unable to accord a high priority.

Finally, pharmacologists should be actively involved in preventive approaches, mainly by participating in the collection, evaluation and dissemination of data as well as by taking an active share in educational activities both for the professional and for the layman.

Research on educational techniques should have a high priority. It is necessary to find out the best educational approaches suitable for various audiences, depending on their general level, professional activities, if any, and appropriate to their possibilities and capabilities to follow the ideas presented during the educational programme.

REFERENCES

Cameron, D.C. Drug Dependence: some research issues. Bull. Wld Hlth Org., 1970, 43, 589-598

Collier, H.O.J. Supersensitivity and dependence. Nature, 220, 228-231, 1968

Collier, H.O.J. Supersensitivity and dependence on cocaine. Nature, 220, 1327-1328, 1968

Eddy, N.B., Halbach, H., Isbell, H. & Seevers, M.H. Drug dependence: its significance and characteristics. Bull. Wld Hlth Org., 1965, 32, 721-733

Eddy, N.B., Martin, W.R. Drug dependence of specific opiate antagonist type. Pharmakopsychiatrie, Neuropsychopharmakologie, 3, 73-82, 1970

Ellingwood, E.H. "Accidental conditioning" with chronic methamphetamine intoxication: Implications for a theory of drug habituation. Psychopharmacologie, 21, 131-138, 1971

Gill, E. W., Paton, W.D.M., & Pertweek, R.G. Preliminary experiments on the chemistry and pharmacology of cannabis. Nature, 228, 134-136, 1970

Gunne, L.M. Personal communication, 1970

Martin, W.B. Pharmacological redundancy as an adaptive mechanism in the central nervous system. Fed. Proc., 29, 13-18, 1970.

Smith, C.B. Effects of narcotic analgesic drugs on brain noradrenergic mechanisms. Paper presented at the 2nd Pharmacological Society Symposium on Agonist and Antagonist Actions of Narcotic Analgesic Drugs, Aberdeen, 12-13 July, 1971.

Wikler, A. Sites and mechanisms of actions of morphine and related drugs in the central nervous system. _Pharmcol. Rev._, _2_, 435-503, 1970

World Health Organization. Treatment and care of drug addicts. Report of a Study Group. _Wld Hlth Org. techn. Rep. Ser._, no. 131, 1957

World Health Organization. Evaluation of dependence-producing drugs. Report of a WHO Scientific Group. _Wld Hlth Org. techn._ Rep. Ser., No. 287, 1964

World Health Organization. Services for the prevention and treatment of dependence on alcohol and other drugs. Fourteenth Report of the WHO Expert Committee on Mental Health. _Wld Hlth Org. techn. Rep. Ser._, No. 363, 1967

World Health Organization. Research in psychopharmacology. Reports of a WHO Scientific Group. _Wld Hlth Org. techn. Rep. Ser._, No. 371, 1967

World Health Organization. WHO Expert Committee on Drug Dependence. Sixteenth Report. _Wld Hlth Org. techn. Rep. Ser._, No. 407, 1969

World Health Organization. WHO Expert Committee on Drug Dependence. Eighteenth Report. _Wld Hlth Org. techn. Rep. Ser._, No. 460, 1970

PROFESSOR JACOB: Thank you very much Dr. Chrusciel. As a pharmacologist I fully agree with your conclusions and I hope these will also be the conclusion of the conference.

We have completed the presentation of the factors related to drug dependence. The floor is now open for discussion.

DISCUSSION

PROFESSOR H. EDERY (Professor of Pharmacology, Israel Academy of Sciences and Humanities): I have been concerned with the marihuana compound for many years. The main point which should be stressed in future research is the choice of the right animal species, in order to make this research worthwhile. I do not think anybody will oppose the view that animal experimentation is important in this field.

In our work, we found that the rhesus monkey is one of the

most useful animals for the study of the effect of marihuana, both in what concerns its mode of action and the search for antagonists. Delta-1-HTC which is one of the active components in marihuana, produces a cataleptoid reaction in the rat, which makes this animal species unsuitable for cannabis research. The sensitivity of the rat is too low and it would require an injection of 10 to 20 mg per kilo body weight, which is a tremendously large dose, in order to produce effects similar to those in human beings. The dog has also been found to develop a cataleptoid reaction, similar to that of the rat. Loren, in the early forties used the ataxia test for marihuana in dogs. However, very large doses were necessary to produce an effect.

In a photograph taken by the police in Israel in one of the places where people smoke hashish, shows a typical picture of pseudoptosis of the eyelids, apathy and loss of awareness of the smokers. The monkey appears to be the only other animal species which responds to marihuana with pseudoptosis. It is probable that this particular response is specific to primates. Dogs injected with delta-1-THC show pseudoptosis and apathy just as in the case of the hashish smokers.

In order to produce this effect in monkeys it is necessary to inject 50 mg per kilo body weight. This is the threshold dose which produces the same effect in human beings. One can conclude therefore that human beings and the rhesus monkey respond similarly to similar doses.

Regarding the second aspect, namely, the search for antagonists, I would like to recall that Russian scientists, two or three years ago, claimed that THC effect can be reversed by the administration of phenitrone. In our hands, phenitrone failed to do so. Similarly, Professor Romax from Los Angeles reported that he was unable to reverse the THC effect with phenitrone. On the other hand the effect of amphetamine is absolutely remarkable. In monkeys, a single dose of amphetamine produces a rapid reversal of the stupor produced by delta-1-THC. I submit therefore that amphetamine is a good marihuana antagonist and that further studies of amphetamine action should be undertaken. Thank you Mr. Chairman.

DR. R. PETZHOLD (International Dental Federation): Mr. Chairman, I have followed with deep interest the papers of Dr. Chrusciel and Professor Deniker relating to the need of additional research in the field we are discussing and especially the need for the development of means for the identification of drug dependents. As a matter of fact, when the International Dental Federation received, in April this year, the invitation to participate at this conference, the Executive Director, Dr. Leatherman of London, and myself wondered on the one hand just how our profession, the dentists, could be concerned with drug dependence and on the other what could we do to fight this new plague in the world. And still, even though scientific investigations touching at the same time the state of the teeth and to drug dependence are numerically few, I am personally acquainted only with the work of Professor Waldir Nesi of Rio de Janeiro on the disastrous effects of marihuana on the lips, the tongue and bucal mucous membrane. There are four aspects of drug dependence which concern directly the dentists. First, the increased incidence of dental caries and periodontal disease found in drug addicts. Statistical evidence for this is still to be established. Second, the possibility of early detection of drug addicts by dentists, since the lack of dental hygiene forces the addict to consult a dentist fairly early. Third, the assistance that a dentist can give to a patient, recognized as a drug dependent. And finally, fourth, the secondary effects whether inhibitive or cumulative, of dental therapy of drug dependents. I am especially thinking of nitrogen protoxide, of anaesthetics and of analgesics administered daily.

It is for this reason, that dentists would like to have at their disposal sure means for the recognition of drug addition, especially in its early stages. The International Dental Federation, through its various organs, such as working groups on dental therapy, could issue recommendations addressed to all the dentists in the world, this of course in accordance with data provided by persons competent in the field, in order to limit the spread of this plague. But this requires that the diagnostic criteria of drug addiction, which ever it may be, be made available to us. Thank you Mr. Chairman.

DR. GELLHORN: I would like to address my question, I believe to Dr. Mendelson or to others as well. In the area of therapy we have heard a plea for additional knowledge in the pharmacology of many of the drugs and this would obviously include their metabolism. Alcohol which is a simple molecule and has been so well known for such a long time, I suppose has had a greater amount of study than any of the compounds about which we have heard today and therefore a great deal is known about its intermediary metabolism. As a result of the studies of the metabolism of alcohol there has been introduced some time ago the compound "antibuse" and in order to inform me better as to whether this type of study of metabolism of the drugs really will prove to be a useful mechanism in the treatment I would like to ask Dr. Mendelson whether "antibuse" is considered by those who deal with the problem of alcoholism, as an effective therapeutic agent and what is the relative advantage of this type of therapy as compared with other forms of therapy.

DR. MENDELSON: The mechanism of action of antibuse has been well known for a long time and in fact the basic aspects of ethanol metabolism are well known. But I must sadly report that studies of ethanol metabolism have contributed more to our knowledge about the kinetics of enzyme function than they have contributed to our understanding of the pathogenesis of alcohol related illness. Succicntly I think one could say at the present time there is nothing that we have learned from the processes of ethanol catabolism in man which would help us first, either uniquely identify an individual who is at high risk for developing an alcohol related problem or enabling us to introduce an effective means of intervention. Antibuse does have eficacious results in certain individuals but it is dependent upon of course the individual's willingness to perpetuate antibuse administration. In many ways antibuse is an ideal drug because our most recent knowledge of this agent indicates that what it does, it blocks the de novo synthesis of those aldehyde dehydrogenases in liver which are involved with the catabolism of acid aldehyde derived from ethanol. So its a good drug, but it doesn't work if it is not used by the individual. I must extend my comments a bit further by going to a statement made in Dr. Chrusciel's paper which I think is a very perceptive and very cogent observation and if I may Dr. Chrusciel quote from your paper : "Agents that block the release of action or that influence the synthesis of biogenic amines should be investigated as a means of preventing the development of drug dependence. Several approaches along these lines have already been undertaken. Research data on the connection between biogenic amines and the action of dependence-producing drugs are increasing exponentially. Most recently Smith (1971) from the NIMH Addiction Research Centre in Lexington has shown that opiates increase synthesis of catecholamines."

Now in the light of this statement I would like to prepose a notion and it goes something like this. That most individuals who use centrally acting drugs use them in my opinion, not to produce the concomitant depressive effects but use them to produce what one might generally call a state of activation - a state of allerting. We know that alcoholics who consume alcohol for a long period of time for example, eventually become more dependent, more despondent, more depressed and eventually develop a great deal more anxiety associated with drug use than they had when they initiated drug use. So we cannot strongly argue that these individuals are consuming alcohol or any other drug to reduce some aversive contingency. Rather, they may be using a drug to produce a state of euphoria or a state of activation or be it that that state is transient; it is brief, it dissipates quickly, but the reinforcing quality of the drug use to initiate that state is very important. Now, what are the concomitants of that state? There again there is some increasing evidence that the activational state if you like has to do with activation in a simplistic way of a number of systems such as adrenergic systems that we believe were depressed. It is quite surprising to examine Dr. Smith's data and other data that have come from the NIMH Addiction Centre and other institutions which show that ingestion of morphine or use of morphine, use of heroin, use of ethanol produces a state of adrenergic activation rather than suppression. Alright - now we can develop a hypotheses. The hypotheses is that if we could somehow block that activating effect we might obviate the effects or the reinforcing qualities that are attendant upon the use of the agent. The next step is what? Well again very simplistically there are a number of places that are working with alpha and beta adrenergic blocking agents - both in experimental animals and in man and initially some of these studies show some rather promising sorts of results. What I am saying is that perhaps we have to entertain some new models concerning the processes of drug action - in terms of what they do to enhance or initiate a change in a feeling state in either an experimental animal or man and not simply go toward a model which is going to block the overall action of the drug such as antibuse or impede its metabolism at some crucial state and produce a dysphoric state in the individual.

PROFESSOR REXED: Since this is a general discussion may I change the topic by a few degrees. This morning's discussion I think, has centered on many aspects of the factors at work in the non-medical uses of drugs and in a general way one could, as of course many people have been saying, discuss these factors within three sectors of operation. The individual, the drug, and the social group - the social setting. It seems to me that the tendency in a group like the one we have before us today is strongly individual-centered and this is a risk. We tend in a group committed mainly to medicine and perhaps to public health to look at the problem as a clinical pharmacological one and as one centering on the individual, his tendency to use drugs, or his fallacy to adopt himself to the risks of drugs if he needs them, and we look at it very individually centered. I would not think one would leave this discussion without reminding people that the area of the social setting is a very important one. Dr. Freedman has touched upon it in so far as he has dealt with the individual in the group of friends and family and the interactions that culture and religion and things like that might have in this setting, but he didn't go further than that. I think if we try to analyse the abuse situation in the world and individual instances in different countries, and don't take into mind the socio economic settings we will not be able to deepen the analyses so far that it would be a great help to authorities and to countries that want to do something against it. I could explain this in several ways. One would be to say, that if you notice in a country that there is large scale heroin misuse intensely concentrated in very slummy areas with very low income groups and very low cultural and social level. Then I think to attack this problem simply by dealing with individuals would not solve the problem to the end. I could illustrate it from my own country. In a recent investigation in the Stockholm area the social authorities took all the registers they had at hand, the registers of the economic situation they had on households, registers of criminal delinquents, registers of work done by the social authorities. All in all, something like ten registers, describing really the socio economic conditions of the population and where they had a possibility to link this with geographical areas. They ran these registers together on a computer and they made a map of Stockholm where the coincidences of all these things were mapped out. It came out very clearly that there were four or five areas of Stockholm where you had a great running together, low income, high delinquency, low cultural situation of the population, etc. etc. All those things which go to point out people who are in a very unfavourable situation in the society and who have been hit one way or other by life and circumstances. I understand quite clearly that it is difficult here to discuss post and propter. You might say that this is the following of something that happened to the individual, it's the effect, or you might say it's a course that you let people be like that. Well, anyway, I think if we don't take up this problem and analyse really what factors in the society are centrifuging people into such areas, and let them stay there, and let society accept the situation that they stay there - and then look at it as simply an individual problem. If we don't attack this problem and try to analyze it then we will get nowhere. In a situation in Stockholm I would say it is now absolutely necessary to go further in this discussion and take up the low income problem, for instance, of these families. The problem of giving these people education and work and at the same time try treatment of them, if they have drug abuse. I forgot to say incidentally, that the frequency of drug joints was closely comparable to the map of these socio economic distressed areas. Here I would only say that this group evidently is not put together in such a way that we could discuss such aspects of the problem and I don't argue for them that we should take them up, but I can understand it is useful sometimes to differentiate problems and to discuss one aspect more than something else. So discussion will probably be useful. But in the future do not forget this aspect of the socio economic analyses of the background factors in society that operate here and don't anyone go home and believe that we can solve the drug abuse problem in any country without taking up these problems for a thorough going analyses. We help to solve such problems as poverty, low educational standards, unemployment, things like that and only if we take these problems sincerely we will get any place on a large scale of drug abuse. My second remark would be for public health purposes and to ask the representative for WHO what could be done with it. I think three things are to be done. Some kind of early warning system studying high risk groups or areas where drug addiction is going on actively to see developments in techniques in using new drugs in taking up new trends of misuse, because as Professor Deniker has pointed out an important part of modern clinical psychopharmacology is done among the addicts and we have to use these for an early warning system. The second need would be to have some permanent system of frequency registration so that we could follow really the trends in the different countries. This afternoon we are going to report from the various quarters on trends and I am sure many interesting things will be said. But the basic fact is that we do not have any sure facts to register as far as frequencies in different countries are concerned. We have more guesses than estimates and a system for random sampling surveys, of some kind, to see trends in frequency of abuse is highly necessary. The third area which I think should be highly valuable from the public health side is to compare treatment and rehabilitation experiments and efforts all around the world. Now these are done on small scale and very randomized everywhere. We do not have any place where they are taken up and compared, evaluated, and this information would be distributed. My question would be - is it possible for things like this to be done? Is it possible for WHO to go in a large scale here? Thankyou.

PROFESSOR JACOB: Dr. Cameron would you answer?

DR. CAMERON: If Professor Rexed would permit, I would prefer to answer this sometime tomorrow as this deals with some of the aspects in my paper.

PROFESSOR RESAT GARAN (Professor of Medicine, Scientific and Technical Research Council of Turkey): Mr. Chairman, I should like to call the attention of the participants to a simple but very instructive observation made more than half a century ago by Professor Mayor, Professor of Pharmacology at the University of Geneva, namely, that the phenonthrenic group of opium alkaloids which produce a state of agitation in the cat (Feline mania), produce equally a physical dependence in man. It seems to me that it would be of interest to confirm this observation and to try to analyze this phenomenon, in the light of recent progress in psychopharmacology, in order to explain the mechanism of production of dependence. Thank you.

DR. FREEDMAN: I am sorry to go from cats and morphine mania to poverty. I did want to respond to two points that Professor Rexed raised. One is, of course, that the actual epidemiology of, especially, narcotic abuse in the United States requires us to take poverty into account. But I have no trouble finding it on our richest college campuses! I have often thought that our Brahmins are as deprived of consequences and meaning in life as our poor. Social disorganization is probably the very important factor in the change of style and motivation of drug taking.

The other point which I think is important (and I thoroughly agree with you about group processes) is that I am just not certain that economic deprivation is as important to contemporary drug abuse in youth as leisure and the ability to buy products for entertainment.

But there are early warning systems that I think are possible, and I think this is terribly interesting and important. I know that some of us professors knew more about contemporary patterns of drug taking (because it began on college campuses) than any narcotic officers. We were in touch with the people who were using the drugs. This "in touch" factor is often missing for us in the Harlem slum, which is terribly disorganized. Patrick Hughes, the University of Chicago Epidemiologist and Psychiatrist, noted one fascinating study which I think is on its way to publication (if it already hasn't been published); namely that a middle class Negro neighborhood that begins to change is at high risk for heroin addiction. Why? Well, in America, the middle class Negro has a hard time finding housing, especially if he is socially climbing. Their neighborhoods turn over about every ten years--new people move in; the middle class are misplaced. The newcomers have no social connections--everybody "responsible" has his eye away from his neighborhood; heroin-taking and crime begin in the back alleys. Instead of neighborhood control and somebody talking about it--nothing happens. Three to six months later the police recognize the fact that there has been an epidemic going on. Now, Hughes and his people documented this as it happened. The type of ability to find out when a neighborhood is in trouble, when a new fad is beginning, is going to require an entirely different spirit of approach than what has characterized the past. I think that kind of epidemiology has a lot to offer.

DR. MENDELSON: I would just like to amplify some of the remarks made by Dr. Freedman in response to Professor Rexed's paper, and that is to differentiate between high risk and high visibility. This is something that I try to get at in my presentation for often we can confuse the two. Before my visit to Geneva, and my trip to Europe, I saw two cases - one alcoholic who had been inbibing very heavily for a number of years and had fallen down the socio-economic ladder and was highly visible to his family and society, but fortunately he had good liver function. Another individual whom I saw was quite intact in his family, social, and economic sphere but was dying of hepatic disease in a hospital room for which he was paying almost $200 a day. Both had a drinking problem, one was highly visible to society the other was not. So although none of us would deny that socio-economic factors, factors of conditions of life are extremely important we should also be aware that those cases that come to our attention, come to our attention as a function of some factors that brings them to our attention. And one factor which brings cases to our attention and problems to our attention are the social and economic context in which they occur. So I would feel that in thinking about global problems of drug disorders whether it be with cannabis - or one involving alcohol and other centrally acting agents we should keep in mind that we see what comes to our doorstep and at times we may with some degree of error make causal associations, that is, because poverty is in a large extent present in a community which we detect a large number of cases of drug problems. It does not necessarily mean that there is a causal relationship between the two and correction of one aspect of the individuals total life style may not mean that we can successfully intervene in what we are thinking about as a primary problem, at least from our point of view namely the drug problem.

DR. EDWARDS: Reading between the lines, I am sure Professor Mendelson himself would tell us that he is not altogether happy with this particular tryptich, and he is unlikely to give it a very honoured place in his own art collection. Division into the social, the biological, the psychological is not an absolutely satisfactory way of looking at things. What we are really so often seeing is the interaction between elements and we need therefore an interactional model of understanding. At a meeting such as this, it is furthermore only a matter of convenience that we focus on the drug problem - a very contrived focus. We are looking at one particular narrowly defined type of behaviour which is socially disapproved, repetitive, and occasionally self-damaging. What we are in the end lacking is not a model of drug dependence but a model of behaviour, and that so far is something which pharmacologists, psychologists and sociologists haven't given us, and that is the real absent guest at the feast. Dr. Mendelson then talks interestingly about adrenergic effects of alcohol taking. How fruitful it might be to turn our vision away from alcohol, transiently, and to look at similar behaviours which are very closely allied in behavioural pattern but ones where no chemical is involved. For instance, gambling, or obsessional hand-washing - there is no externally administered rewarding chemical re-inforcing these behaviours, and we might be able to look at basic processes of dependence in much purer culture. Dr. Mendelson remarks on case finding were of much interest. I would suspect that in my own country there is rather little evidence that alcoholism is commoner in any particular social class. The condition does however manifest itself in many different ways, and society itself determines what will be called a "case". Thus, if we are going to talk about case-finding, we have to ask "What do we mean by a case?" Another point - case finding to what purpose? What is the point in designing new methods to detect more drug addicts, more alcoholics, when we don't know what to do with the cases which we have already detected? Lastly, when I sit among pharmacologists I wonder why I don't do it more often, but then go on to wonder whether multidisciplinary research between widely different disciplines really is possible. If you start off with a room full of men and women representing very diverse disciplines, do they all in the end just get depressed, and fail to understand one another?

DR. EVANG: I would just like to state that the relationship between the economic, political and social factors that increase the amount of abuse is a complex one. We should not be fooled by the fact that the richer the country in the western world, the more abuse is found and draw the simple conclusion that the more money you have, the more drugs you buy. It may be that the frustration thresholds in those societies where there is very much money floating around is of another type than in a more primitive society. Also, as a parallel, may I remind you, that if you tried to establish your figures on the prevalence and incidence of a specific disease based on hospital patients only - which is in itself a very dubious affair - then certainly your result would depend upon the type of hospital service you have got in that country. If we have an all comprising hospital system where there is no economic deterrent from getting into the hospital, you get a completely different picture from that in a country in which you have to buy your way into a hospital, or where certain groups, rich or poor, perhaps have a preference. These are very complicated studies. I think we should never fall into the trap of blaming one another for not giving enough emphasis to the facts that we think are important, but we should try - as Dr. Edwards said - in this cross professional group to evaluate them between us.

PROFESSOR REXED: I do not quite agree that drug abuse is especially prevalent in the rich countries. As a matter of fact there is evidence that poor, low standard countries are subject to a great deal of drug abuse. And within these countries, one finds a delineation of misuse groups which are closely structured to social levels. Recent investigations in my country have also shown that there is a higher prevalence of alcoholism in the lower social structure. However, as I said I do not want to fall into the trap of changing post into propter, but I am quite clear that causations are difficult to establish. Dr. Mendelson you were yourself falling into the trap when you described to us the first instance. Since you yourself belong to a high cultured, highly educated, high income professional group, you find it rather distasteful to admitting an explanation showing that just because you have permitted individuals to be otherwise this might be a factor of drug abuse.

PROFESSOR JACOB: I was asked to summarize the presentations that were themselves brief summaries of the subject. My task is made still more difficult since ours is a multidisciplinary conference and we often use the same terms to express different concepts. I will therefore ask you to consider my own bias, as a pharmacologist.

At the beginning of his excellent presentation, Dr. Chrusciel, recalled that drug abuse is often compared to a communicable disease, the chemical substance being the agent. Dr. Isbell rightly insisted on the dose relationship of drug dependence, this would correspond to the size of the infection. He also differentiated between the various types of dependence and between the various types of dependence producing drugs. Drugs should be differentiated as it is done with the agents of disease; nevertheless, they are all iatrogenic substances - having as a common meeting point precisely their ability to induce dependence.

Just as in the case of infectious diseases, the evolution of dependence relates to the susceptibility of the individual. The biological factors involved are as yet poorly understood. That such factors exist, is, in my opinion, not a matter of doubt since there are interspecific, intraspecific and individual susceptibilities in laboratory animals.

All the same, this communicable disease presents singular pecularities. The causal agent does not reproduce in the body of the host as is the case of an infective micro-organism. It is the host who often inoculates himself with the agent for the first time and then proceeds to reinoculate himself. There exist models in animal experimentation where various techniques of auto-administrations produced very interesting results. However, if these models prove the development of an appetite for various types of drugs they are still very inadequate for the study of other important factors related to the susceptibility in man. Professor Deniker has identified two such factors. To begin with, at least in France, drug dependents are psychopaths; secondly, they are characterized at least during a period of their affection, by a state of adynamism which has therapeutic consequences. These two factors are of certain importance in the therapeutic perspective, as was well demonstrated by Professor Deniker. I am certain they will be discussed further during the conference.

Finally, the environment, just as in the case of infectious diseases, is of primordial importance. But again, the problem here is complicated since the environment not only favours or otherwise the initial innoculation but also influences the possibilities of the reinoculations necessary for the development and maintenance of drug dependence.

The epidemological study of Mr. Shoham has identified some of the sociocultural factors, especially factors that appear not to increase, but rather to decrease the susceptibilities of certain communities. This is just as important as the attempt to identify factors favouring dependence. It represents the reverse side of the coin, and the understanding of the phenomenon cannot be satisfactory unless favourable and unfavourable factors are each made the object of study, rather than be content with reciprocal indirect deductions. This principle is being currently applied in pharmacological research the interpretation of which, because it is less complex, is less amendable to error when both positive and negative controls are used.

Dr. Mendelson insisted on the importance of a methology designed to avoid, in the socio cultural field, too hasty generalizations or too simplistic reifications. In doing that, he underlined one of the major aspects of the problem facing us, namely, the search for new methodologies. This is valid in all fields, beginning with the field of animal pharmacology in which for example, new techniques for the rapid development of dependence states in small animal species could lead to more efficient screening of potential dependence producing drugs as well as the establishment of better biochemical correlations.

Much research is necessary in the field of pharmacology, as detailed by Dr. Chrusciel as well as in the clinical and therapeutic aspects of drug dependence, a subject so well covered by Professor Deniker. I wish to congratulate the speakers for the excellent presentations which incited such useful discussions. Each speaker accomplished the extraordinary performance of presenting within a very short time, the essentials of the difficult problem in which he is competent.

THIRD SESSION

CURRENT PATTERNS OF ABUSE OF DEPENDENCE PRODUCING DRUGS

MODERATOR: Dr. Daniel X. Freedman

DR. FREEDMAN: I am sorry that Dr. Kusevic is not here at the moment, so we will begin this session which is on "Current Patterns of Abuse of Dependence Producing Drugs" with a discussion of patterns in France by Dr. Mabileau.

THE PATTERNS OF DRUG ABUSE IN FRANCE

Dr. Jean F. Mabileau

Inspector-General, Social Affairs Inspection Branch

Ministry of Public Health and Population, Paris

As late as 1965 the representative of France, in describing the situation, was still able to say that "addiction did not present social problems in France, but was still an individual evil which called for close attention. No country could be sure of being exempt from the scourge. In France, drug addiction was mainly of therapeutic origin and, thanks to the strict regulations and to the cooperation of doctors, the number of cases could be kept quite small..."(Report of the Twentieth Session of the Commission on Narcotic Drugs, E/4140, E/CN. 7/488, p. 31).

The following year, in December 1966, he was able to report that - "although there were relatively few drug addicts in his country supervision was nevertheless active..." (Report of the Twenty-first Session of the Commission on Narcotic Drugs, E/4294, E/CN, 7/501, p. 29).

During the last four years, however, the situation has changed and side by side with the classic cases of drug addiction of therapeutic origin, cases of many different types have been noted, often involving dependence on more than one drug in young people.

While the position is not as sombre as in some other countries, it must be treated with due seriousness and the author will now attempt to describe its salient features.

DEVELOPMENT OF THE SITUATION

It is always very difficult to establish correct statistics. The main sources of information are:

The police departments which are under the control of the Ministry of the Interior, the Directorate-General of the National Police;

the National Gendarmerie;

the Customs Department;

the psychiatric hospitals;

the inspectorates under the control of the Ministry of Public Health and Social Security, particularly the Pharmaceutical Inspectors of the Health Department.

The Narcotics Bureau of the Ministry of Health, which provides the secretariat for the Interministerial Narcotics Commission, has brought together all the information on the subject.

It should be noted that while the phenomenon gives the impression of having assumed epidemic proportions since summer 1968, there has been a parallel increase, and a considerable one, in the number of officials in the departments involved in combating it. In one sector, there has been a rise from 40, mainly concerned with breaking up international drug trafficking rings, to 300 police officers spread all over the country and able to question a greater number of offenders, whether addicts themselves or traffickers.

ACTIVITIES OF THE SUPPRESSION AUTHORITIES

According to the Ministry of the Interior the number of summons from 1965 to 1969 was as follows:

Year	Addicts	Traffickers	Total
1965	60	105	165
1966	82	86	168
1967	155	120	275
1968	207	154	365
1969	994	206	1200

Breakdown of the number of arrests by drug involved, age and a distinction-addicts, traffickers

	Opium		Morph-ine		Heroin		Cocain		LSD		Canna-bis		Var. pharm. products		Cate-gory Totals		GRAND TOTAL
	T	A	T	A	T	A	T	A	T	A	T	A	T	A	T	A	
19 yrs & below		5	2	2	42	87			3	30	48	254	5	31	100	409	509
20-34 yrs	11	9	7	8	37	156			23	39	173	682	11	37	312	931	1243
35-49 yrs	3		4		16	4				1	23	21	5	2	51	28	79
50 yrs & over	4		4		12	2					2	4	3		24	6	30
Arrests	17	14	17	10	157	249			26	70	246	961	24	70	487	1374	
TOTALS	31		27		406				96		1207		94		1861		1861

During 1970, from 1 January to 31 December, the authorities dealt with 913 cases and made 1861 arrests or summons, involving 1374 addicts (A) and 487 traffickers (T) or middlemen.

During the year the following amounts were seized:

Opium	11822 g
Morphine base	248610 g
Heroin	77765 g
Cocaine	2 g
LSD	3544 doses
Cannabis	645748 g

In 1970 the courts passed 677 sentences, divided as follows:

Prison sentences of over 2 years	33
Prison sentences of 1-2 years	42
Prison sentences of 6 months - 1 year	64
Prison sentences of one week to 6 months	401
Prison sentences of less than one week and fines	71
Discharges and no cases to answer	66

DISTRIBUTION OF KNOWN ADDICTS BY SEX AND AGE-GROUP

On the basis of information from various sources brought together by the Narcotics Bureau of the Ministry of Public Health and Social Security, statistics have been drawn up for 1970 covering 730 addicts and 826 instances of the use of addiction-producing drugs by them:

			Total		Percentage	
Age Group	Men	Women	Simple	Cumul.	Simple	Cumul.
15-19 years	191	83	274	-	37.52	-
20-24 years	259	66	325	599	44.52	82.04
25-34 years	64	17	81	680	11.10	93.14
35-44 years	2	2	4	684	0.55	93.69
44-49 years	3	0	3	687	0.41	94.10
50 years & over	7	1	8	695	1.10	95.20
Not indicated	31	4	35	730	4.8	100.00
	557	173	730	-	100.00	-
	or: 76.3%	or: 23.7%				

A comparative study has been made of utilization in 1969 and 1970, covering 643 and 826 instances of drug consumption respectively.

	1969	1970
Cannabis	50%	53.4%
Heroin	16.6%	21.2%
Amphetamines	9%	6.5%
LSD	2.2%	5.4%
Morphine & Opium derivatives	9.8%	5.2%
Synthetic narcotic drugs	2.5%	2.7%
Barbiturates	2.3%	1%
Phenmetrazine	6%	0%
Sundry	1.6%	4.6%

The relative importance of cannabis consumption is noteworthy. It is being consumed more and more in the shape of cannabis resin (hashish).

A significant increase in heroin-taking is also perceptible and has become obvious from study of recent addicts. There has also been a decrease in the prices of heroin in the two regions most affected (the South of France and the Paris region) and there have been several seizures of relatively important amounts of that drug, adulterated to varying degress, intended to supply the local demand.

The statistics also indicate a recrudescence of abuse of LSD and although the cases are few in number they tended to increase in 1970.

Abuse of amphetamines and similar products, taken by mouth at first and then by intravenous injection of crushed tablets dissolved in water, led the authorities to take very strict measures of regulation which have succeeded in reducing the size of the problem considerably, particularly in respect of phenmetrazine (Preludin).

ABUSE OF PAREGORIC MIXTURE

In 1969 cases of abuse of paregoric had been reported. Young addicts evaporated it and injected the opium-containing residue after adding water.

In December 1969, therefore, the conditions of sale without prescription were amended. The preparation on unrestricted sale is now made up of equal weights of a sugar-rich syrup and paregoric. Abuse has stopped because the preparation is no longer of interest to addicts who can no longer inject themselves with it after evaporation.

Paregoric by itself, on medical prescription, is still giving rise to a few cases of abuse, but markedly fewer than before. Pharmaceutical chemists are indeed very wary when prescriptions for the preparation are presented to them and inform the Pharmaceutical Inspectors of obviously forged prescriptions or prescriptions for amounts which seem to them excessive.

The restrictions on the supply of paregoric led addicts to search for other opium-containing products sold off prescription, such as tablets or suppositories containing very small amounts of powdered opium or poppy extract, and to subject them to various procedures. This situation has forced the authorities to withdraw the existing exemptions and to place the previously exempted substances in the Table C class of poisonous substances (issued against prescription), a solution which is of no hindrance to sick persons requiring genuine treatment.

LEGISLATIVE MEASURES

The penal legislation was extensively reshaped by the Act of 31 December 1970.

As early as 1969 the public authorities, alerted by the increase in the consumption of narcotics, had begun to investigate how the law needed to be reformed. The purpose in view was to look for a solution in the case of drug addiction which would not necessarily be of a penal nature and on the other hand to adapt better to the needs of today the laws and means of control relating to the fight against drug abuse and drug trafficking.

The new law introduces some important innovations in the penal and social treatment of drug addiction, a scale of penalties distinguishing sharply between the addict and the trafficker and a strengthening of the means of suppressing illicit traffic.

Thus while the mere consumption of narcotic drugs by an

individual may attract a penalty, the judicial authorities have the right to determine whether a prosecution would in fact be advisable taking into account whether the drug-taking has been a repeated offence and whether the addict will participate actively in the medical and social measures of treatment prescribed for him.

As for traffickers who have participated in the importation, manufacture or export of narcotic drugs, they are henceforth liable to ten to twenty years' imprisonment combined with security measures. This sentence can be doubled if the offence is repeated.

Finally, the period of detention without being charged in the case of infractions of the laws on narcotic drugs can be as much as four days and the right of search and seizure has been extended and reinforced.

CONCLUSIONS

The public authorities are particularly alert in face of a situation which has developed over the last few years and is marked by an increase in the number of cases of drug abuse, relatively frequent consumption of cannabis and a rise in heroin addictions.

These cases of drug abuse involve young, sometimes very young, people. The market seems to be in process of being organized in some regions with a view to systematically obtaining more and more customers from more and more varied socio-economic groups.

Psychosociological research has been undertaken by the National Institute of Health and Medical Research with a view to gaining a better knowledge of the possibilities of taking preventive measures of a primary or secondary nature.

The Act of 31 December, 1970 should enable effective control of the scourge of drug addiction.

DR. FREEDMAN: Many thanks, Dr. Mabileau. I am sure that later we will have time to discuss this interesting battle of wits between the scientific establishment and the drug takers--a kind of benign chemical warfare--paradoxically, I found that an interesting device. Our next speaker will be reporting from Scandinavia.

PATTERNS OF DRUG ABUSE IN SWEDEN

Professor Bror Rexed

Director-General, National Board of Health and

and Welfare, Sweden

Though the following remarks apply mainly to the situation in Sweden, some mention will be made of other Scandinavian countries. There are, amongst the participants at this conference persons who are more knowledgeable than I about conditions in their own countries and may wish therefore to complement my presentation.

Just as in the case of France, it is the 1960's that have marked a change in the patterns of drug abuse in the Scandinavian countries. As far as Sweden is concerned, and this probably holds true for Denmark, it was around 1965 that the authorities discovered that we were facing a large misuse of central nervous system stimulants. The discovery came as a shock, even though the situation had started earlier.

There is one conclusion that can be drawn from our history. You may have a rather widespread misuse without realizing that you have it. I suspect this is a common experience. At the moment the pattern seems to be something like this: There is quite a wide exposure to drugs all over Scandinavia, perhaps mostly in Sweden and Denmark. Figures from school children in the upper classes and from conscripts in Sweden show that before they reach the age of 20 - probably something like 50% of the young people have been asked whether they would like to take narcotics or psychotropic drugs, in one or the other form, either by friends or by pushers or by somebody else. About 15 to 25% of them, different in different regions and groups, have experimented with, one or more drugs. Less than 5% have experienced repeatedly. One to two percent depending on the kind of drug and the area, have continued to take drugs permanently. The lesson to be drawn from these statistics is as Professor Freedman said, that there is no reason to be pannicky just because we know that there is a large scale exposure. The important thing is whether we have been able to inform and to create a preventive attitude among people so that they do not react to this possibility. The situation just now is that the experimentation and misuse of cannabis is slowly spreading all over Scandinavia, it is no doubt most frequent in Sweden and Denmark. I do not know in which country it could be said to be most frequent. The wave of misuse of central stimulants that we had in Sweden reached a peak around 1969. It has now subsided. There is no doubt about that, but it is still entrenched in the big cities, in Stockholm, Gothenburg, and Malmo and it is my feeling that the real source of misuse is in the criminal or half criminal and adjacent circles of the population. Other groups are also recruited from time to time. We cannot say exactly how many misusers we have but if we follow the illegal traffic and the price in the market it is clear that misuse is widespread in certain groups, although it is limited. It is probably less than it was in 1969. This was brought about by a rather integrated approach, intensified police action, intensified customs action but also a large information and educational campaign, setting up treatment centres in hospitals, intensified work in the social rehabilitation area, and a campaign to improve information to young people through the schools.

The misuse of central stimulants in Denmark is perhaps not so widespread as in Sweden, perhaps it never has been. Norway and Finland,seem to have a lesser problem than the other countries and no doubt we will hear more about it and about the reasons. They may be geographic, cultural or there may be other explanations. A Commission in Finland not long ago concluded that the problem of hypnotics, sedatives and tranquillizers was a bigger one than the problem of narcotics. In this connection one should add that also in Sweden, this problem is a considerable one and not to be forgotten.

Among other drugs, we see a disquietening increase in the use of opiates. This is new. We had seen it before in what you might call health related cases, either professionals or patients - but now we see it in Stockholm and to some extent in

Gothenburg. Young people have gone directly to opium mainly morphine base remarkably enough - taken by injection. They manage to dissolve it in a simple way and inject it. We have between two and three hundred cases of young people turning into misuse of this kind of opiate, directly without any other experience. We have had so far very few cases of heroin. As for LSD there is experimentation and perhaps a little increase over earlier periods, but nothing really important to speak of.

There is one final point I would like to make and that not all Scandinavian countries have the same pattern of drug misuse. The reasons are difficult to explain. One possible reason is the difference in tourist traffic, especially amongst young people.

The source of drugs of the central nervous system stimulants type is undoubtedly in Southern Europe and the Mediterranean area where they are produced in legal factories. These substances seem to be able to slip through the chain of commercial activities and reach the misusers. In our experience the misuse of these drugs is dangerous. We had hundreds of deaths amongst the misusers. Furthermore the habit takes people rapidly down the social scale and out of any kind of ordered life.

THE PATTERNS OF DRUG ABUSE IN THE UNITED KINGDOM

P. Beedle

Head of Drugs Branch, Home Office, London

**

Public recognition of drug abuse as a community problem within the United Kingdom is barely 10 years old. Public interest, first aroused in 1963 by spree use of amphetamine tablets by young people in London, was heightened in 1965 by a sharp rise in heroin addiction and sensational publicity about LSD. The "drug problem" has since passed through several well-publicised phases without any clear threads or relationships. Following is a summary of the main threads detected from a wide variety of sources:-

general and special reports by the police;
routine analyses of prosecutions, conviction and seizures;
Analyses of prescriptions and legitimate supplies of drugs;
statutory notifications of narcotic addicts;
analyses of hospital admissions and surveys of other population groups;
reports by export committees of enquiry;
published and unpublished research.

HEROIN AND METHADONE

Before 1960 non-therapeutic addiction to narcotics in the United Kingdom was negligible. In 1960 known heroin addicts numbered 94; only one was under 20 years of age. In 1967 however, the picture changed radically; out of 1929 known narcotic addicts; 1299 were taking heroin and 381 of these were under 20. This rapid increase originated in a small cult among jazz musicians using morphine and heroin stolen from a hospital. It gained momentum from the emergence of a new type of addicts - young, disturbed, often aggressive and delinquent - and from the over-prescribing of heroin by a few doctors in London from whom the addicts sought treatment.

Early in 1968 measures were taken to set up centres for in-and out-patient treatment of heroin addicts, to confine to specially licensed doctors the right to supply heroin to addicts at these centres, and to require all doctors to notify narcotic addicts. Since then the rise in numbers of known narcotic addicts has been halted; and fewer addicts under 20 years of age have been reported. During 1968 some 2782 addicts came to notice; during 1970 the number was 2661. At end 1968, 1746 addicts (764 under 20) were known to be receiving drugs; at end 1970 the total was 1430 (405 under 20). A number of special treatment centres have used methadone maintenance in their programmes. This is reflected in the analysis of 1430 addicts at end 1970:-

992 addicts were receiving methadone (of whom 254 were also receiving heroin);

183 addicts were receiving heroin (either alone or in combination with drugs other than methadone);

255 addicts were receiving drugs other than methadone or heroin and their addiction was mostly of therapeutic origin.

Most methadone and heroin addicts continue to be concentrated in London, with a few very small groups in other English cities. Over one-third are in employment.

STIMULANTS

Oral abuse of amphetamines by young people in the early 1960s was closely associated with a vogue in London for - 'beat clubs' and 'discotheques' , all night dances and week-end parties. The vogue spread gradually to smaller towns and rural areas and rather slowly in the North of England, Scotland and Wales. In 1964 penal sanctions were introduced for unauthorized possession of amphetamine preparations. The annual number of convictions rose from 958 in 1965 to 3603 in 1969. Towards the end of 1969 police forces began to report signs of a decline in abuse of amphetamines and this trend was also indicated in surveys of young people admitted to London remand homes. The number of convictions fell in 1970 to 3128. This development may be connected with recent initiatives by doctors in many areas to eliminate or cut down the use of amphetamines in medical practice and with changes in youth culture generally.

Until 1967 intravenous abuse of amphetamines was uncommon. When heroin was restricted early in 1968 a few London doctors began to prescribe methamphetamine in large quantities and a number of heroin addicts quickly formed an intensive cult within the London area. When in October 1968 supply of methamphetamine was restricted by voluntary agreement to hospital pharmacies, these doctors took to prescribing amphetamine sulphate powder for injection until pharmacists ceased to dispense their prescriptions a few weeks later. For a time the increasing trend to intravenous abuse continued with a variety of substances but in the past year this phenomenum has been declining.

CANNABIS

All the available evidence points to a general growth of cannabis use over the last 20 years and to rapidly increasing spread both geographically and socially in the past five years. Total convictions for cannabis offences in the years 1963-70 were:-

1963	1964	1965	1966	1967	1968	1969	1970
663	544	626	1119	2393	3071	4683	7250

The drug appears to be most popular as an intoxicant among the young and to be used socially among small groups. Little, however, is known about the total extent or current patterns of use at particular age or social levels or about the inter-relation of cannabis use with abuse of other drugs.

LSD

The vogue for this drug developed widely in 1966 and 1967 and then declined throughout 1968 and 1969. In 1970 the police reported widespread and increasing use throughout the country and convictions for offences (mainly of unauthorised possession) numbered 757 as compared with 159 in 1969. There are no reliable data on prevalence of use but various surveys have indicated that abuse of LSD is now part of the general drug-taking scene among young people under 25 from all social backgrounds.

SEDATIVES

Dependence on barbiturates is not uncommon among the older sections of the population. Abuse of barbiturates by young people has so far seemed insignificant. In 1970 some 300 London addicts resorted to intravenous abuse of short-acting barbiturates with particularly damaging results; this practice has not spread. During 1968 and 1969 there was substantial abuse of a proprietary hypnotic compounded of diphenhydramine and methaqualone. Penal sanctions were introduced in 1970 and this spree phenomenum seems to have declined of late.

GENERAL

Three more general features perhaps deserve mention.

(i) There is no evidence that the rapid rise in heroin addiction in the 16-20 year age group in 1960-1967 was derived from widespread school age cultures of drug abuse.

(ii) Sample surveys of the school age population have suggested that the numbers of those with experience of drug taking is small and that most of those with such an experience give up taking drugs after a few experiments.

(iii) These and other surveys have suggested that by contrast with the early 1960s both the large majority of young persons who experiment with drug taking and the minority of persons who have established a habit of drug abuse no longer confine their interest to one or two drugs but are ready to use a wide variety of substances.

** This paper was read by F. Stewart, Drugs Branch, Home Office, London.

PATTERNS OF DRUG ABUSE IN IRAN

Professor K. W. Bash

Psychiatrische Universitäts-klinik, Berne

Drug dependence in Iran means almost exclusively dependence on opium and its derivatives, i.e. "shireh", a decoction from opium residues, and heroin. Morphine is relatively little used, which may be attributed at least in part to intensive campaigns against morphine addiction instituted some fifteen years ago.

Before entering on the topic of dependence, it is well to recall that there has long been, and despite legislation still is, a considerable body of opium users who have not sunk to dependency and are unlikely to do so - people who smoke their several pellets a day as a European may take his glass or two of wine without degenerating into an alcoholic. Most opium is smoked, not seldom as a social ritual in male society. This circumstance is of some importance for understanding patterns of addiction, because the use of opium in itself carries no social stigma.

The vast majority of drug dependents in Iran were and perhaps still are poor or stand on the lower fringe of the small middle class and turn to drugs not for deliberately cultivated enjoyment, but for the relief of distress. There are some indications however, on which I shall touch later, that the relations are changing. It has been claimed that, rhoughly in descending order of magnitude, the distress is likely to be medical, economic, social and psychological, though the boundaries between these groups are admittedly unsharp, several causes are likely to work together and the psychological factors are in my opinion certainly more important than a number of non-psychiatric workers in the field have presumed them to be. Azaraksh (1) found that among 1167 addicts hospitalized in the first years after the prohibition of opium in 1956 41% had become addicted because of physical complaints. This was a mainly urban population (Tehran). I should guess that the proportion of such addicts in the countryside would be higher. Up to the institution of the Health Corps about 1964/65, medical services outside a few cities were extremely scanty. Over wide rural areas self-medication, if any, was inevitable, and there was only one drug that effectively relieved bodily pain. Since the ailment itself rarely was treated, the anodyne was employed in ever increasing amounts until dependency on the drug for its own sake supervened. The proportion of "medical" addicts is likely to diminish substantially as medical services rapidly improve and spread.

To the observer who has known the country before and since the reforms of 1962 there can be no question that there has been general economic improvement. Nevertheless, poverty remains, and many a poor man seeks forgetfulness of his cares in the drug. He is likely to utilize the cheaper, but more toxic shireh, and not seldom he must turn to some form of crime to get it. The prohibitive price of illicit opium too has helped to limit its use by the lower economic groups.

Social factors and patterns of dependence are too complicated to be analyzed in a presentation of this length. Rather than dwelling upon ancient injustices, which are sufficiently known, I would refer to the enormous social upheavels which Iran as a swiftly developing country is undergoing. They can be achieved peacefully, but not painlessly. Old values, both material and spiritual, disintegrate; new ones are at best nebulous except for an often overpowering demand to possess all the supposed "good things" of technological civilizations now, at once and all together. Faced by the sheer impossibility of such craving, persons of both low and high degree take substitute satisfaction in drugs and, in their frustration, may gallop into dependence. The pattern is as familiar in Western as in Iranian youth.

A long known pattern of dependency is the use of drugs to counteract work fatigue, more newly to compensate for job satiation and frustration. I must fear that, as industrialization spreads in Iran, this pattern will expand, continued intensive application to a very narrowly circumscribed, unvarying task being altogether foreign and repellent to flexible habits of labour conditioned upon agriculture and handicraft and cherished over centuries.

The social and the psychological patterns of dependency in Iran mix, and far too little attention has been paid to the latter. These increase in importance with rising socio-economic level, so that the middle and upper levels probably contribute more than their proportionate share of addicts. A study by the Tehran School of Social Work (3) on a stratified sample of 30 hospitalized addicts concluded that most had negative feelings toward society, were unhappy, had psychological problems going back to childhood, desired things which it was impossible to achieve in Iranian society, experienced family breakdowns and were encouraged to use drugs by false stories.

To look, maybe, slightly deeper into the problem, let me review in a few words the relevant materials from the psychiatric epidemiological survey of Shiraz which my wife and I conducted for the WHO in 1969/70 (2). We had under scrutiny a random sample composed of 930 persons over six years, drawn from two strata, a poor and a well-to-do neighborhood. We found seven cases of unambiguous drug dependency, all male, five relating to opiates, one to alcohol, and one to both. Three, including the alcoholic could be classified as psychopaths, one as neurotic. One had begun taking opium as an overworked scullery boy for the relief of gastric pain and insomnia. Two were so far demented and abandoned that no useful information regarding the origin of their dependency could be procured; both had been quite well-to-do and were now ruined. Notable seems that only one, the scullery boy, came from a definitely poor family, the other four being from the middle class.

As is generally known, Iran in 1956 prohibited the cultivation of the opium poppy and all traffic in or use of opiates except for medical purposes. In the early 1960s informed guesses by the most experienced officials of the health and narcotics control services about the number of persistent addicts ranged between two and four hundred thousand. Under the pressure of massive smuggling from adjacent countries Iran changed the law in 1969 to permit opium cultivation and production under license and to allow all addicts over the age of 60 to apply for cards entitling them to receive 15 grams per day of opium produced under government control at a moderate price. Persons below 60 considered to be in need of opium because of chronic illness or irremediable addiction may receive opium under similar terms upon a physician's prescription, approved by the local health authorities. At the same time drastic penalties were introduced for peddling and smuggling, including the death penalty for large scale dealers. The number of addicts registered at the end of 1970 was officially reported to be 80,000 (4). Informed persons suggest that the true number of addicts is likely to be the double of this. Practically all physicians and enforcement officials agree that it is still too early to assess with any reliability the effect of the new policies, though several of my acquaintances have cautiously mentioned signs of improvement.

Heroin, not encountered in our Shiraz sample, is a relatively new, but menacing and still increasing problem in Iran. This holds especially for the Tehran metropolis, whereas from at least one province (Fars) a "dramatic drop" is reported since heavy penalties were introduced in 1969 (5). Knowledgeable persons from the United States have told me that heroin addiction originated there in the slums and was regarded as a symptom of poverty and malaise. As far as my observation goes, and I was living in Iran when heroin addiction began to manifest itself there in the 1960's, the Iranian wave took the opposite course, arising among the youth of well-to-do families and students, perhaps also as a symptom of cultural desorientation, and spreading downward.

In conclusion, a word as to other drug dependencies. The role of hashish was, and I believe still is, insignificant. In 1960 alcoholism was almost unknown except among foreigners. It is increasing, but is still a minor problem and, except in rather small westernized circles, does not enjoy the social sanction that the use of opium does. Alcohol is little employed for its taste and quality, but more or less frankly for inebriation. The many pharmaceutical hypnotics and analgesics that so often lead to dependency in Europe and America cannot compete with opium in effectiveness or even in price, costing about double as much in Iran as in Europe, and therefore only exceptionally pose a problem.

LITERATURE

1. Azarakhsh, H.A. : Drug Addiction in Iran. Proceedings of the First Symposium on Addiction in Iran, ed. by M. Mahloudji. Shiraz 1963.

2. Bash, K. W., and J. Bash-Liechti: Psychiatrische Epidemiologie in Iran. Sitzungsberichte der Deutschen Gesellschaft für Psychiatrie und Nervenheilkunde. Zentralblatt für die gesamte Psychiatrie 201 (1971), 255-256.

3. Fort, J.: Draft Report to WHO; 1963.

4. Government of Iran: Report to the United Nations for the Calendar year 1970 on the working of the International Treaties on Narcotic Drugs.

5. Moharreri, M.R.: Some Aspects of Drug Addiction in Iran. Report to WHO Expert Committee 1971 (unpublished).

DR. FREEDMAN: Thank you so much Professor Bash. I don't know what you observed when you were in Chicago. I think the American pattern of opiate abuse, as you know, involved American Civil War veterans, middle class ladies consuming vegetable tonics, liberally laced with alcohol and opium for making one feel a bit better. When heroin was introduced this was mostly used by the middle classes. I guess it was between 1914 and 1920 that our poverty groups began to become involved. There was a big epidemic after the Second World War, but that epidemiology deserves a closer look.

Our next speaker is Dr. Dube from India. Dr. Leong Hon Koon will follow with a report from South East Asia.

PATTERNS OF DRUG ABUSE IN INDIA

Dr. K. C. Dube, MBBS, DPM (London) FAPA, FAMS

Professor of Psychiatry and Superintendent

Mental Hospital, AGRA-2, India

The use or abuse of dependence producing drugs in India has long traditional and social roots going back to 2000 BC to early Indo-Aryan civilization. The current use should be viewed in the historical perspective to understand the religious, social, cultural and other influence that play a part in it. In "Charak Samhita" (a book on Ayurvedic medicine by Charak in vedic literature), the whole of the Chapter 24 is devoted to the qualities and ill effects of the excess use of alcohol. The administration of alcohol as an anaesthetic in labour was practised in ancient India.

Cannabis use was described in India in 800 B.C. Its habitual use as a common intoxicant exists since 1000 A.D.[1] First recorded instance of opium cultivation in India is found in 15th century. Since then the use became quite prevalent during the Moghul period of history. Serving of opium was a universal custom among the Rajputs and in court functions. Later, opium smoking became fairly popular in North India. It was also used by working mothers for administration to infants and children to keep them quiet - a custom which now hardly exists. The oral consumption of opium was stopped by the Government of India in March 1969 and a rationed dose is sold only to registered addicts. The consumption of this drug (through excise issue) has fallen from 256938 Kgs. in 1957-48 to 2791 Kgs. in 1960-61.[2] Opium preparations used for smoking are usually in the form of chandu and madak, smoked through special contrivances.

The use of cocaine has existed for the last 75 years. During the first two world wars, cocaine use was much more prevalent specially in North India where people chewed it with betel leaves as it was believed to increase sexual powers. Due to restrictions and scarcity, its use now is uncommon.

The concern with medical and social problems initiated the Government of India to appoint two Enquiry Commissions in 1883-1895 and 1954-55. When the historic U.N. Convention of 1961 declared a number of drugs illegal subject to penalties, India voiced her reservation in relation to Cannabis resin.

After India gained her independence, prohibition for alcohol was introduced in several states. The sale of opium and ganja was also restricted by a system of licensing. In some states it is totally banned. The sale of charas is banned all over India. In spite of social change and stir for prohibition, hemp preparations can easily be procured. This confirms the fear expressed by India in 1961 at the United Nations' Convention to be true. The Drug Controller reveals that the use of barbiturates and Dexedrine had gone up by 5 times within the last 5 years. The barbiturate habit is common in those states where the State Governments have given recognition to "Unani" and Ayurvedic practitioners e.g. in Punjab. Several unqualified practitioners misuse authority leading to mal-practice and leakage of drugs. Some of the unscrupulous practitioners habituate patients to the use of opiates. The use of injectable preparations is frequently resorted to by medical men themselves.

Although, various intoxicants have been in common use, the habit is indulged in only by the pleasure seekers and the addicted. It never had the blessings of religion, yet the drugs developed some religious and social importance. The intellectuals, however, considered it a custom decrepid society. The orthodox Indian Society always frowned upon the habitual indulgents, whether rich or poor. The use of hemp, specially the smoking of ganja and charas was confined to the lower social strata of society. While higher caste people never indulged in the use of alcohol, the use of bhang had some amount of social tolerance and at times approval. During social functions e.g. marriage celebrations and festive occasions like Holi festival (festival of colours), the use of bhang is uninhibited and unrestricted. Even the indulgence by children is overlooked.

The use of cannabis is mostly confined to North India but not uniformly. There are places of high drug use and low drug use. The former are mostly confined to places of pilgrimage like Mathura, Varanasi and Puri where there is high concentration of users among whom Sadhus and Brahmins predominate. Most so-called "Sadhus" indulge in the habit of ganja smoking. Among such communities like "Chaubey" Brahmins of Mathura, the use of bhang (dried leaves of cannabis) is universal. Among the villagers ganja and charas are usually smoked through "chilam" (Indian earthen-ware pipe) mixed with tobacco. Bhang is used in the form of a drink or a pill to swallow. Some times confectionary is also used (Majoon).

Of late a new trend in the pattern of drug use is emerging. Alcohol is freely used. The more easily procurable drugs, i.e. ganja and charas are becoming quite popular and considered a fashionable habit. The trend is further increasing with the influx of large number of foreign hippies and drug traffickers who have invaded most cities. The vice is spreading fast among the student community and the youth in general. A recent study reveals that in the campus of Delhi University 50% of male and 8% of the female students have had a drug experience at one time or the other. The total number of addicts in college campus is estimated to be about 200 while nearly 5000 rely on regular thrill seekers for occasional drug ration. The drug cult has touched 1/20th of the student population of Delhi. Another study of 150 male and 100 female students of Delhi colleges disclosed that 1 in 8 has visited the drug dens on and off the campus. Admission to these drug dens is not easy. A fresher has to stand around for some time before he is admitted. Pot orgies are common. Bus-stop sessions are not a rare sight around the campus; example: A student waits in the crowd for a bus restlessly, he looks here and there and hurriedly pulls out a packet from the secret pocket of his jacket, empties out a black tablet on the palm of his hand, shakes the tobacco from a cigarette on it and crumples the tablet mixing it with the tobacco. The powder is rolled in a cigarette paper and he lights it. He takes a deep puff known as "dam" and passes it on to his companion who is gleefully waiting for the "mall".

The present extent of drug use is quite in contrast to a couple of years ago when only about 2% of the male students could be classified as drug addicts, while the number of girls indulging in the habit was nil.

A matter of serious concern is that the drug epidemic is increasing, affecting both boys and girls. The habit is more common among the students with better schooling and good family status and is more prevalent in fashionable cities. "Fag" and "Booze" is a thing of the past. Sipping wine and smoking the "pot" is a new trend. Sleeping and anti-sleeping pills, tranquillizers, aspirine, etc. are adjuncts used for "tuning in". Stronger drugs like heroin, cocaine, morphine taken by intravenous injection (mainlining) is not unusual. LSD has been introduced in Delhi and a number of students are using it.

There are no proper statistics available on the incidence of drug addiction in India. There are difficulties in obtaining comprehensive information such as: (a) absence of proper agencies to collect such information; (b) large size of the country and different patterns of drug use in different regions (c) heterogenous social customs; (d) due to ambivalent religious, moral and social attitudes and sanctions, the drug users prefer not to divulge their addiction.

In a large epidemiological study of mental morbidity in and around Agra, which is a low drug use area, it was found that among 16725 adults, 381 persons were habitual users. In the same study there were 26 drug users among 607 mentally ill giving a rate of 6.81% as against the rate of 3.56 among the non-users - the difference being significant at 1% level. The percentage of mentally ill who use more than one drug was 16.25%. The users of cannabis (bhang and ganja) comprised 10.67%, while the percentage of alcohol users was 2.2 only. The rate of 9.84% among psychotic users was conspicuous as against the rate of 2.21% among the normal individuals. The psychoneurotics had a low rate of 1.81% only. Among the manic depressive psychotics, the rate was highest, being 16.22% while in schizophrenics it was 13.61%.

Recent histories of the patients admitted to Mental Hospital, Agra, are scanned. Among the 358 consecutive male admission, 100 patients were found to be inhabitual users of cannabis preparations. This indicates that one in every 3.5 psychotics admitted gave a history of drug use.

The figures collected by Ministry of Labour and Employment in 50 different centres spread all over the States, revealed that 10 to 20% of the working class men are given to drinking, and about 15% of the working class families use alcohol. 15% of the adult males are given to drinking. Within the last 10 years, in Delhi alone the consumption of liquor has almost doubled., In the State of Uttar Pradesh cannabis use was the highest, while the consumption in 1956-57 for the whole country was 1.22 kg, the consumption for the same period in Uttar Pradesh was 2.185 Kgs.per thousand persons. Since then the consumption has increased manifold.

These indicators point out the alarming trend towards which the youth of India is drifting. It is, therefore, necessary to have a proper agency for conducting a nationwide survey for assessing the extent and magnitude of the problem. Such a venture would be of immense help at national and local levels, to take effective steps to tackle this social evil.

---

1. The cannabis sativa grows wild almost anywhere in India but is found mostly in the foothills of Himalaya. The drug is used in the form of Bhang, ganja and charas (sulpha). Bhang consists of the dried leaves of the female plant. It is ground into a paste and mixed with various flavouring agents. It is taken as a sweetened drink with milk or else used in the form of small pills. Ganja is the dried flowering top of the plant and after mixing with tobacco is smoked through a chilum (earthenware pipe). Charas is a resin obtained from the leaves and flowering tops and is also smoked. Sometimes a confectionary of cannabis called majoom is used.

2. Opium is cultivated in several places under strict government supervision in Central India. Bihar, U.P. and Panjab, raw opium is eaten. Opium is smoked in the form of 'Chandu' and 'Madak'. 'Madak' is prepared by boiling opium with charred acacia leaves and the mixture is allowed to thicken and rolled into small pills and smoked through a pipe or reed. Chandu is another preparation obtained from boiling opium in a copper vessel. The thick scum having the appearance of coal tar is called 'Chandu' which is smoked through a pipe.

PATTERNS OF DRUG ABUSE IN SOUTH EAST ASIA

Dr. Leong Hon Koon

Medical Officer in Charge, Outpatient Services,

Singapore

South East Asia may, for the purpose of this very brief survey of patterns of drug dependence, include:

Hong Kong and Macao
Burma, Thailand, Laos, Cambodia, Vietnam including "Yunnan" and that "NO MAN'S LAND" where opium is grown;
Singapore, Malaysia (Malaya, Sarawak and Brunei)
Indonesia (Sumatra) and
the Philippines

This region covers -

(i) the area where "Yunnan" opium grows, the main source of the illicit opium in the East. The term "Yunnan" is smugglers' jargon for the geographical area lying in the region of the trans-Salween states of Burma, Southern Yunnan, Laos and north-west Thailand;

(ii) countries where opium-smoking still remains - Malaysia, Singapore, Burma;

(iii) countries where heroin dependence has taken root and become greater problems than opium smoking - Hong Kong and Thailand;

(iv) countries with an American "presence" where onto the age-old opium pattern is imposed, a new pattern of GI heroin dependence with the imminent likelihood that it will overspill back into the local population - Vietnam, Laos and Cambodia.

With regard to dependence of the marihuana type this region includes countries where cannabis is grown and smuggled into neighbouring countries - Indonesia, Sumatra.

Up to 1941, the smoking opium was legal in Singapore and Malaya. Addicts were registered and could buy opium from government chandu (opium) retail shops. In 1941, there were 16,552 addicts on the Singapore registers. After World War II, the smoking of opium became illegal and punishable by fine or imprisonment.

The average opium addict was a Chinese male, aged 49 years who had smoked opium for about 18 years, a labourer or unskilled worker. He took opium for relaxation or for pleasure, to relieve pain, to allay fatigue and other symptoms and for its imagined properties. The susceptibility of the individual, the addicting nature of opium, its availability and the force of tradition were factors in the aetiology of his dependence (Leong 1959). He first smoked prepared opium, then swallowed it; later, in straightened circumstances, took dross. He would take 3.2 packets of prepared opium a day. He spent about 3/8 of his daily income on drugs, taking one packet of 11.7 grains or 0.756 grammes costing $1.20 (Singapore), three times a day.

This was the typical addict of the fifties and the sixties. This pattern of drug dependence, that is, the smoking of opium among mainly the Chinese, a left-over of a government-permitted system, carried over the Japanese occupation of South East Asia (1942 to 1945) prevailed in Malaya, Singapore, Hong Kong and Burma. In Thailand, too, the pattern was similar.

After World War II, the patterns for Hong Kong and Singapore took divergent paths.

During the 1950s, for a multiplicity of reasons, Hong Kong experienced a change over from opium to heroin.

In Thailand too there was a change towards heroin. Only in Singapore and Malaya was the pattern still that of opium smoking.

However, in Singapore, along with the opium-smoking addicts, there had always existed small coteries of morphine injection addicts. These in the beginning were associated with certain areas in the city and with certain particular racial groups and occupations and even in certain streets. Thus pockets of morphine (injection) dependence were found among the Henghua, the Hokchia and the Hockchew, in the Muar Road, Ophir Road, Johore Road areas, among the trishaw-peddling, labourer community.

The typical morphine addict was generally a decade younger than the opium addict and he would have been first an ordinary opium addict smoking 3 packets of opium (costing $3.60 Straits) a day. One day after coming in contact with a morphine addict or a morphine den operator, he would be persuaded to switch to morphine with the persuasion that it was cheaper, and that it would render him less liable to be arrested for possession of opium smoking utensils. He would be started with a subcutaneous dose of 20 cents an injection which would substitute for the smoking of $1.00 worth of smoking opium. Soon however, he would be stepping up the dosage and after a few months, also switch to intra-muscular and later intra-venous injection.

Thus in Hong Kong opiate dependence would be 80% heroin, 20% opium with negligible morphine injection; in Singapore it was roughly 80% opium, 20% morphine with nil heroin.

The reasons why Singapore is free from heroin addiction are not known; one could only speculate that the factors which led to heroin addiction in Hong Kong, viz. overcrowding, a large refugee population, an influx of "operators" from China, relative availability of "chemists" who could be persuaded to cooperate with the operators in manufacturing the heroin from morphine were not present in Singapore.

MARIHUANA DEPENDENCE

As with the opiates, there has been a change of pattern in marihuana dependence; whereas however, in opiate dependence the change was in the drug from opium to morphine, with marihuana the change has been in the persons affected from the old-timers to a new milieu. In Singapore, from the earliest times, ganja, the local name for marihuana, has been associated with Indians and Pakistanis who brought over the traditional habit from their own motherland; and its use has been confined to a small number of these particular groups. As such, it was a very localised problem, with a "ghetto" like character and did not give much cause for concern.

Ganja, was smoked as a cigarette in "rokok daun" (a locally obtained dried leaf used as tobacco paper) in an ordinary cigarette or in a pipe, plain with a chillum (a locally obtained cone shaped, clay pipe) or more elaborately with the local version of the hookah-josah or nargileh.

Around mid-1964 owing to confrontation between Malaysia (of which Singapore then formed a part) and Indonesia the smuggled supplies from Indonesia were cut off, and traffickers introduced the seeds to local farmers. The addicts mostly seamen who became unemployed introduced the drug to young "pleasure seekers" who frequented the "sarabat" stall (a roadside coffee stall) in the backlanes off Jalan Besar, Tanjong Pagar, Geyland, Bukit Timah and even Katong. There were even reports that ganja cigarettes (the newspapers referred to them in the American style, as reefers) were being offered to teenagers for smoking sessions to the accompaniment of music and to cabaret girls in their flats for the entertainment of their customers and to rock and roll music groups. It was noted that ganja taking had spread among the Malays in certain districts.

Around 1969, it was estimated that the addict (ganja) population was approximately 1, 500 comprising about 500 "regular" addicts and about 1, 000 "unaccustomed" addicts who took it for kicks (Singapore's 1967 estimated population was 1, 955, 600).

The 1969 Singapore Year Book reported "an enhanced marginal interest in 'POT' shown by the younger groups."

Interwoven with these patterns, the writer is seeing, at a hospital drug dependence clinic with colleagues and a medical social worker, young persons from middle class families, English-educated, who are taking ganja and "pills" in the form of methaqualone. Two particular brands were seen, one was a Roussel Laboratories products, Mandrax, the tablet, marked with letters MX, became known as MX pill.

It would be interesting to speculate and we can at this stage only speculate why of all available pills in the market it should be these two particular brands. The MX pill and the M pill. Both of these thought to contain morphine. Perhaps the mysterious M with the unknown X had something to do with it. The third unknown pill is so mysterious, it is called No Name Pill, while a fourth, even more mysterious is known as the American pill, black in colour.

I have, in speaking with patients about these pills, the impression that they imbue them with qualities which in fact they do not quite possess.

As for alcohol dependence, Khoo and Fernandez (1971) reported that the problem exists, and gave a brief profile of the Singapore addict, and gave the incidence as ranging from 300 to 1,300 per 100,000 at risk.

The above patterns in broad lines present the picture of dependence in South East Asia; the patterns are not static and much remains for fuller study.

REFERENCES AND SOURCES

1. Glatt, M.M. & Leong, H.K. (1961) Alcohol Addiction in England and Opium addiction in Singapore: some differences and similarities. Psychiatric Quarterly, 35:1.

2. Khoo, O. T. & Fernandez P. (1971) The Problem of Alcoholism in Singapore. Singapore Medical Journal 12. 154.

3. Leong H.K. (1959) Opium Addiction in Singapore Dissertation for D. P. H. London.

4. Hong Kong Government. The Problem of Narcotic Use in Hong Kong: A White Paper laid before legislative Council - Nov. 1959 (Hong Kong Govt. Printer 1959).

5. Straits Times, New Nation, Sunday Times, Sunday Mail. Current newspapers in Singapore.

DISCUSSION

DR. FREEDMAN: As I read these papers, I was simply fascinated to get some authenticated feeling of what other countries were experiencing at this time; but I find it very difficult to generalize from what we have heard. I am not so certain that it would not be useful for those who have reported before the end of the meeting to think of what it is that they see. Or perhaps all of us students ought to be asked how our views have changed. Change seems to be the rule; even in what seems to be a somewhat tranquil culture of Iran. Also, the youth, educated youth, seem to have a very special transnational culture, as if the rock music serves the same function as perhaps the ancient temples did when people gathered there for special occasions. It is interesting to see who experiments with something new, with something apart from the ordinary, and how they transmit it. We get some feel of that from some of these reports, and I think we probably need more of a feel of it. I wonder, Dr Edwards, if you would like to comment on how this strikes an epidemiological mind.

DR. EDWARDS: How it strikes this particular epidemiological mind is that he has learnt a lot from his friends and wishes the discussion could go on for several days. I wish however that it could be turned into a workshop rather than a round table conference, so that we could look in detail at the actual way in which the fascinating data that we have all listened to this afternoon has been gathered, and examine together some of the basic scientific questions which underlie these endeavours. What has interested me is, for instance, the number of correlates of drug use which have been shown us. This theme has been illustrated so interestingly by Professor Rexed, and we must all want to see his map of Stockholm. Correlates of cannabis use and psychosis have been reported from India and the correlates of class and addiction from Iran. This whole business of the interpretation of correlation is one of the central questions of epidemiological research, and I would like to exemplify this by quoting an example from England, an unimportant country situated somewhere between Wales and the sea. I would quote for this purpose research data which at first really seems to support Professor Rexed's view to the hilt. If you go to London you will find that the problem of chronic drunkenness arrest is circumscribed rather sharply to certain city areas and is largely a matter effecting the least skilled and privileged classes. Then you go to Alcoholics Anonymous and you find that alcoholism is not obviously correlated with city area, and you find A.A. membership heavily biased towards the managerial and professional classes. You go to hospital admissions statistics, to cirrhosis death rate figure, and you get a variety of conflicting answers. What then does the original "city areas" correlation mean? I think that one would have to be careful in how "causally" one interpreted the crude finding, and it turns out indeed that most of the men being arrested in these particular sectors of London, come originally from Ireland and Scotland. There are then too, of course, classical studies of schizophrenia to show that schizophrenia is concentrated in certain city areas. I do not know what our friends who are geneticists would tell us about how they square that sort of result with their genetic findings - anyone familiar with the schizophrenia literature would again warn us about too simple interpretation of a "city area" correlation. It would however be a great mistake to take a polar view and to say that living circumstances are never important, and I certainly have no wish to set up such a polarity. I believe that there is no doubt as to the correlation, but how you interpret that correlation, I don't know. I think there is too a parallel here to Jack Mendelson's very important question - you will find that there are certain sorts of family constellations related to genesis of drinking pathology, what is it in that family? What is the intervening process? Not enough toys? Not enough mothering? By analogy, if you find that a particular census tract is a breeding ground - or a migration ground - for drugs or alcoholism, we must usually admit that we are totally mystified as to the processes at work. Whether it is not enough dollars, whether it is not enough hot or cold water, or whether it is....I don't know. But I would like to see sometime this whole question of the scientific assumptions underlying this type of epidemiological research taken to a high level.

PROFESSOR REXED: Mr. Chairman I would agree with Dr. Edwards that we should be very cautious in explaining correlations. May I just cite, to show the importance of the social situation of people for the kind of delinquency that we are discussing, an expression of a man, Dr. Gustav Junsen in Sweden who investigated a large number of delinquent children and he put as a motto in his book, the following: "the British say that it takes three generations to produce a gentleman - It takes four generations to produce a delinquent."

DR. DUBE: I was very interested to hear a report from Professor Bash that alcoholism was not much known in Iran before 1960. All the same "Shiraz wine" is famous and known from times memorial.

PROFESSOR BASH: Mr. Chairman, I reply briefly to Professor Dube who has justly commented that the wine of Shiraz is famous and has pointed out the paradox between my findings and its literary reputation. I have come gradually to the conclusion that the Shiraz wine has been more read and written about than drunk, and I would support this by a remark that would also be a note of caution. In the Sufi philosophical literature and poetry, wine (sherab) and drunkenness (mastee) may mean many things and are not always to be taken literally. They may connote any form of ecstasy or any means of reaching it with or without drugs.

DR. EVANG: I have one question relating to Dr. Bash's statement on the first page of his paper - second paragraph ending with these words "because the use of opium in itself carried no social stigma." Now this is an important point. We are told time and time again that 'mores' are shifting. I remember that we asked once, the distinguished colleague who is also present here today, Dr. Leong Hon Koon, who has extensive experience from South East Asia, in a similar group in Geneva about this, and he commented approximately as follows: "In the Western world I understand that a father can say to his son - 'let's go to the bar and have a drink together' - in my part of the world a father who is using opium regularly could never say to his son - 'my son, let's go to the opium den and have a smoke together'." This is one point and I would ask this question to Dr. Leong again. I would agree with Dr. Edwards, that we are missing badly sound epidemiological studies in this field. Why is that so? Not because the epidemiological method is not developed, not because we do not have people willing to do the work and money to finance them, but because some of the factors which you have to deal with are very difficult to identify. The critical epidemiologist - and God bless him, I will say - wants things that he can define, and unless he can describe the factors unambiguously he is not enthusiastic including them. I am referring now to the social factors, the economic, political, philosophical, etc. We refer glibly to a number of such factors in the multiple causation of drug dependence, for example in young people, but what do we really know? Take the five nordic countries. They are very close to one another in so far as social conditions, 'mores', philosophy, and so on, are concerned. Nevertheless, they developed five different patterns of drug use and drug abuse - why is this? Is it because two of the countries who may now be more similar than others in their drug abuse patterns, are more strongly influenced by another, dominating country in the world? Or is it for other reasons? Now these are the factors that are obviously important and where unfortunately the epidemiologist is facing great problems. Our next problem is, and I would provoke Edwards perhaps to speak on that tomorrow, to what extent have we here to supplement the epidemiological research model in its traditional sense with other forms of research and to what extent are they acceptable with their merits? I am speaking of, for example the sociologists. Could we educate them into a more epidemiological approach - could we come on better speaking terms with them?

DR. FREEDMAN: I never thought this conference would come to the question of the educatibility of the sociologist. It is going to be very interesting!

DR. MABILEAU: Mr. Chairman, mention was made of the family relationship in relation to opium. I should like to give an example. Having directed a United Nations mission in the State of Shan, namely, that part of Burma west of the Salouen River, between Salouen and the Chinese border, I visited quite a number of opium dens. It was customary at that time in 1964, that fathers brought along their son, their young son, not for the purpose of making him smoke but for him to prepare the opium pipes. One of the natural conclusions by our mission was to demand - in any case their dens are now closed according to information from the Burmese government - that the custom of allowing children into the dens should be immediately discontinued. These children, evidently, took up the habit of opium smoking under the influence of the paternal example.

DR. FREEDMAN: Before we close, I too would like to echo something that I heard going around the table. Parenthetically, I am struck at the phrase, "......of the desire for things that are impossible to achieve," that Dr. Bash was speaking of. If you apply it to the psychology of adolescence, you can see their age specific vulnerability to drugs to satisfy such longing or striving. In other age groups, probably things run differently. I used to joke with the students in America that perhaps these drugs were fine for people over 30, because they were so boring and dull and settled, and nothing could shake them up, not even LSD. I don't think, after a conference like this, I can make such statements--it has been stimulating. But I would say that the points that should be eventually elaborated are what have been referred to as intervening processes. I feel that social psychology as a field has simply missed the boat. I don't know whether there are many sociologists--social psychologists--here. We need some sort of scapegoat! But I think that it is their business to begin to try--really as participant observers, as cultural anthropologists--to begin to record really "how do people learn from each other?" This is what we are trying to understand. I am sure there is a gold mine in everything we have heard in terms of that kind of process.

DR. GELLHORN: Thank you Dr. Freedman. I should like in closing todays session to speak for the audience in terms of thanking the participants for a very informative, very provocative day. At one point during the course of the day I had the feeling that if Voltaire had been here - he would have said appropriately as he did "that doctors prescribe drugs of which they know little, for diseases of which they no less into humans of which they know nothing." But I have the feeling that at the end of the day we are beginning better to define the issues and I am hopeful that there will be greater opportunity tomorrow to have fuller discussion such that both the area of the 'knowns' and the 'unknowns' can be better defined and some of the aspects of the interdisciplinary research and collaboration which has been so eloquently spoken today can be portrayed more effectively.

DR. FREEDMAN: We are going to continue the topic of the patterns of dependence producing drugs and the first speaker will be Dr. Andréasson who will speak on the dependence producing drugs and traffic safety.

DR. ANDREASSON: Mr. President, Ladies and Gentlemen. On behalf of the Executive Committee of the International Association for Accident and Traffic Medicine, I thank you Dr. Btesh for having taken up on the programme the problem of dependence producing drugs and the traffic safety. After having listened to the discussion which took place yesterday, I think this paper which I am going to present is beside the main point of the Conference. On the other hand, it is an important part of traffic medicine.

DEPENDENCE PRODUCING DRUGS AND THE TRAFFIC SAFETY

Dr. Rune Andréasson and Professor Roger Bonnichsen

Swedish Government Laboratory for Forensic Chemistry

The risk of a driver under the influence of alcohol being caught by the police is generally considered to be very small. It might happen through a report to the police from some persons seeing him drinking, a police raid or a minor road accident where the smell of alcohol on the driver's breath betrays his condition or the driver attracts the attention of a police patrol by driving in a way that makes the police suspicious. Instances of drug misuse are generally detected in this way. It means, in effect, that a drug taker is unlikely to be caught unless he is so heavily under the influence that he cannot drive properly or that he shows obvious signs of intoxication when stopped in a police spot-check.

In cases of accidents or when the driver is killed, it is seldom that an analysis for drugs is made. Even if an analysis has been carried out it is usually difficult to draw definite conclusions about the significance of the result. There are so many factors to be considered beyond the individual tolerance to the drug and the time during which the patient has been using the drug concerned. That is why we know so little about the role played by drugs in road safety and why so few drug users are detected. We know much more about alcohol and its effect upon human behaviour. Besides, it is much easier to detect alcohol intoxication since the smell of alcohol in the breath clearly reveals its consumption and the police have been especially trained to detect symptoms of intoxication due to alcohol. It must however, be emphasized that with low blood alcohol levels the symptoms of alchol impairment are not always obvious and therefore many persons seem to be sober although they have consumed so much alcohol that their ability to drive must be considered impaired.

We have made a study of the relationship between blood alcohol levels and the degree of intoxication, based upon the results of a medical examination carried out at the same time as the blood sample was taken. We found a very poor correlation. It was not until the blood alcohol level reached 0.1 per cent that even an experienced physician could tell whether the patient was intoxicated or not. If the subject is a habitual drinker, it is not unusual that even such blood alcohol levels as 0.2 per cent are found by chemical analysis yet the doctor's verdict is "not intoxicated" or "slightly impaired". This has been confirmed by other observers. The result of these studies and our knowledge of the effect of alcohol on driving ability makes us very sceptical of legislation based only upon a clinical examination of the driver. This is true also for cases of drug intoxication. Furthermore in some psychiatric disorders the symptoms of the illness itself may resemble intoxication.

We believe therefore that the results of blood and urine analysis must be obtained in order to bring a person to court for driving under the influence of alcohol or drugs or both. Concerning drugs, however, much more research is still needed to be able to correlate with any certainty the result of the analysis with the degree of intoxication.

Although it is difficult under practical police working conditions to calculate the exact blood alcohol level from breath tests, many types of apparatus have proved useful. In countries where alcohol consumption is legal, it is of great importance that the police be equipped to estimate the blood alcohol level on the spot so as to avoid subjecting innocent people to a medical examination and a blood test, in accordance with the law.

Numerous articles have been published on the subjects of alcohol and road safety and the dangers of alcohol have been statistically demonstrated by many scientists. When it comes to drugs we have no statistics nor do we know much about the role played by them in traffic accidents.

Scientists differ in their opinions of the seriousness of

the drug problem. Some dismiss it as being of minor importance, while others exaggerate the danger of drugs to road users, both drivers and pedestrians. We know, however, that about 15 to 20 per cent of all drivers detained by the police for drunken driving have taken a medicine of some kind in addition to alcohol; medicine that in various ways, impairs driving skill. Some of these drugs will have an additional effect, others will intensify the effect of alcohol.

The use of narcotics by young people and their use of central nervous system stimulants (CNS) differs in many respects from the usual drug misuse and will not be dealt with here, although it is not an uninteresting problem from the viewpoint of road safety. Table I shows some figures from the medical examination of CNS misusers arrested by the police while driving a car.

Table I. Intoxicated by Central Nervous System Stimulants (CNS) Reports accompanying the blood and urine samples sent to the Government Laboratory for Forensic Chemistry in the beginning of 1971.

Sex	Age	Consequence of the accident	Conclusion of the medical examination	Amount of CNS
Male	22	?	Intoxicated	0.1 mg (blood)
				(urine)
"	27	Car theft	Restless, unsteady movements, agitated tremor	15 mg
"	26	Driving without licence	"	80 mg
"	26	Bad driving	"	17 mg
"	36	Car in defective condition	"	94 mg
Female	30	?	"	30 mg
Male	26	Collission	Not influenced	33 mg
"	48	Driving without licence, careless and reckless driving	?	130 mg

In most countries tons of various therapeutic drugs are sold, but fortunately many such drugs only occupy space in the medicine cupboards and are never used. The number and kind of drugs one can buy without a prescription varies from country to country. So far we do not know of any research to find out if drugs are misused to a greater extent in countries where the sale is free than in countries with strict regulation of such agents as amphetamine and phenmetralin. It is however definitely known that the illegal narcotic trade starts from countries that allow a more or less free sale of certain narcotic drugs. Control of the sale of narcotics and the measures to be taken to stop it are thus international questions.

As is the case with alcohol the body gradually develops a tolerance for most drugs with the result that the user must take larger and larger doses to get the desired effect. Unfortunately drugs can remain for a time in the body, thus resulting in an accumulation, which will take a long time, after abstinence, before all the drug has left the body.

Table 2 gives some figures from the medical examination of some drivers who have taken methaqualon.

Table 2 Cases with methaqualon sent to the Government Laboratory for Forensic Chemistry, January-March, 1971.

Sex	Age	Consequence of accident	Conclusion of med. examination	Meth. per 100 ml. of blood
Female	34	Collision	Unsteady walking	0.4 mg
Male	40	Driving on the left side of the road	" "	0.4 "
"	18	Car theft, car damaged	" "	0.3 "
"	24	Reported by police	" " (addict)	0.4 "
Feamle	46	?	Unsteady walking (slurred speech used to tablets)	0.7 "
"	24	Car damaged	Unsteady walking	0.5 "
"	46	Over ditch bank	Almost unconscious	0.3 "
Male	42	Accident	Unconscious	0.3 " (+0.06% alcohol)

The following case history is of interest: A person was slightly injured in an accident. He was prescribed Tikanox (which contains among other things acetylsalicylic acid and mebumal) and Doleron for his pains. He had previously been treated for depression. A couple of weeks after the accident he again went to hospital for his depression and was prescribed large doses of psychopharmaceuticals. He was not warned against driving his car and nobody asked whether he was taking other drugs. As a result he was soon afterwards involved in a traffic accident and was accused of driving under the influence of drugs.

It is thus no longer sufficient for a physician to consider only the somatic damage caused by the drug. He must consider his prescription from a broader perspective and must pay attention to the effect his prescription plays on the daily life of the patient. The physician must also constantly watch and control the consumption and make sure that the patient does not continue to use drugs prescribed by another physician. It should become just as common to analyze the patient's urine for drugs as it is for sugar or protein. Otherwise we run the risk of getting as many drug misusers as we have alcoholics. One must not forget that it is not infrequently that a physician starts a drug misuse. Sooner or later there must come a reaction from society against the steadily increasing drug consumption. In fact some control could be achieved by a central registration of prescribed drugs, suspected of being used as narcotics or having a habit forming effect.

Swedish law allows drivers to take as much alcohol as is consistent with a level of 0.05 per cent in their blood. It means that the law lets the individual judge whether this amount of alcohol influences his driving ability in any way. The responsibility is his own. We can probably take the same view with regard to therapeutic drugs, at least when the patient has become accustomed to the drug and becomes acquainted with its effects.

In Sweden the National Board of Health and Welfare has printed a brochure that describes the danger of drugs to traffic safety and seriously warns against mixing alcohol with medicinal drugs. This brochure is intended for distribution by dispensing chemists to persons who buy the drugs.

In order to prevent traffic accidents due to alcohol intoxication and drug misuse, information about the effect of the medicine should be given by doctors to all patients for whom they prescribe a drug that may influence their driving skill. Such information to alcoholics and drug misusers is likely to have very little effect; many of them do not even have a driving licence-it has either been cancelled or they have never been able to get one. Thus perhaps measures against getting hold of a car without licence could prove as valuable as information.

Since alcoholism and drug misuse are regarded as illnesses in Sweden a new bill proposes to give offenders medical treatment in addition to imposing a fine.

DR. FREEDMAN: Thank you very much Dr. Andréasson. We will now hear from Dr. Wan-Yu Chen.

DR. WAN-YU CHEN (Professor of Medicine, National Taiwan University): I would like to say a few words on the incidence of acute alcohol intoxication in Taipei and its relation to traffic accidents. From the beginning of the year 1967 and through August 1971, 2575 cases of acute drug intoxication were admitted to the National Taiwan University Hospital following road accidents. Of these 299 or 11.8% were due to acute intoxication with ethyl alcohol, 6 to methyl alcohol which resulted in 2 deaths, and 15 to chronic alcoholism. We were able to carry out blood tests on 95 of the 299 cases of acute ethyl alcohol intoxication. The results are set out in the following table.

Alcohol Blood Levels at time of Accident

Alcohol level	Male	Female	Total
Below 50 mg %	9	4	13
50-99 "	11	11	22
100-149 "	12	11	23
150-199 "	12	4	26
200-249 "	7	7	14
250-299 "	3	0	3
Over 300 "	3	1	4
Total:	57	38	95

As the table shows, there is a 3 to 2 male predominance. 72% of the cases were aged 20 to 50 years. Of the group of 95 cases, four died as a result of injuries during the accident.

DR. FREEDMAN: Thank you Dr. Wan-Yu Chen. Before we hear a presentation of Dr. Cameron's paper, Dr. Fernand Goffioul from the Royal Academy of Medicine in Belgium has requested the floor.

DR. F. GOFFIOUL (Royal Academy of Medicine in Belgium): Mr. Chairman, Gentlemen, I shall speak to you briefly on the practical repercussions of drug dependence on the restoration of withdrawn driving licences. The considerable danger to life in respect of road safety caused by the abuse of psychotropic drugs is not generally well-known and secondary prevention is badly organized. The lack of knowledge arises from several causes: attention is concentrated on the apparent, momentary condition at the time of an accident, on the clouding of consciousness and psychomotor deterioration and not on the lasting state of simple, mixed or composite psychic dependence on one or more psychotropic substances and alcohol or non-medicamentous substances. We would like early psychiatric evaluation at the time of preventive detention or the first intervention of the law. It seems to us eminently desirable. This evaluation would necessarily involve complete and properly oriented psychological, biological and social examination. As a result it would be possible to see whether one or several psychotropic substances had been habitually, regularly or accidentally taken in a manner affecting the driving of a vehicle. The main outline of the basic personality and its possible connection with drug dependence would show whether the patient was neurotic, psychotic or psychopathic. When the right to drive has been taken away and some degree of drug dependence or alcoholism has been blamed or could legitimately be suspected, a driving licence should only be granted again if it is proved that the drug dependence has been effectively controlled, the proof, if possible, being given by a specialized team. Repeated biological measurements, the regularity and quality of the reaction to psychotherapy and information on social factors over a sufficiently lengthy period of time would provide a body of usable evidence. This procedure would not require fresh legislation in many countries but merely a revival of a spirit of cooperation between the judicial authorities and psychiatrists specializing in this subject. Such cooperation should be established at the time of the first judicial proceedings with a view to evaluating the case and drawing up a programme of treatment. The psychotechnical centres responsible for the examination preceding the restoration of driving licences should be particularly alive to this problem. A period of two years without relapse is a firm indication of solid improvement. In order to avoid prejudice to professional activities and more generally so as not to create a feeling of frustration which might lead to secondary offences, provisional licences issued for a limited time could be granted when psychological, biological and social findings were favourable but the period of time elapsed insufficient for a final judgement. Secondary prevention carried out in this way would afford better protection for the individual and society not merely from the point of view of road safety but also from that of work safety. Our experience is the basis for our resolves and they are expressed in the conditional because medicine, like strategy, is the art of the possible and you all know that the art is a difficult one.

DR. FREEDMAN: Thank you Dr. Goffioul. Dr. Cameron please.

PATTERNS OF ABUSE OF DEPENDENCE PRODUCING DRUGS-SOME RESEARCH APPROACHES AND STRATEGIES

Dale C. Cameron, M.D., M.P.H.

Chief Medical Officer, Drug Dependence

World Health Organization, Geneva

INTRODUCTION

Drug-taking is the subject of much concerned discussion in many parts of the world and the stimulus for a variety of activities designed to reduce the present and future extent of such behaviour. Among the factors that may be involved in the recent upsurge of interest in this field are the use of certain dependence-producing drugs by persons traditionally thought not to be involved with their use in a given country, e.g. heroin and central nervous system stimulants by middle - and upper-class youth in some Western countries. Another factor may be the invasion of certain areas popularly associated with the use of a particular intoxicant by other intoxicants, e.g. the use of cannabis in countries that have traditionally been oriented to alcohol or opium and *vice versa*. Also involved may be the increasing ease and rapidity with which ideas, things and persons may be transported from one place to another.

Since its inception, the World Health Organization has been concerned with the identification of dependence-producing drugs and other aspects of drug dependence. Its Expert Committees and the World Health Assemblies have been in the forefront in inviting attention to the problems associated with particular drugs including alcohol. Meeting in May 1970, the Twenty-third World Health Assembly adopted a broad resolution on drug dependence that makes the following points, among others:

"Noting with great concern the extensive and serious public health problems resulting from the self-administration of dependence-producing drugs, in particular among the younger age-group;

".................

"Convinced that there is an urgent need at local, national and international levels to implement improved preventive, treatment and rehabilitation measures;

"1. RECOMMENDS that the World Health Organization should encourage and assist the development of improved preventive, treatment and rehabilitation programmes and the pursuit of needed knowledge in the field of drug dependence;

"2. URGES Member States to promote preventive, therapeutic, rehabilitative and research programmes in this field; and

"3. REQUESTS the Director-General:

> (i) to develop means for the international collection and exchange of data on the prevalence and incidence of drug dependence, and on the human and environmental factors associated therewith....".[1]

The following year, the Twenty-fourth World Health Assembly, among other things, underscored the importance of such activities and approved the programme expansion proposed by the Director-General, "especially the collection and exchange of data, the analysis of all medical, social, cultural and economic factors contributing to drug dependence, the conduct of research and training programmes, and the evaluation of existing programmes and the recommendation of new programmes".[2]

The following comments on research needs and strategies stem in large measure from the two World Health Assembly resolutions cited.

---

1 Off. Rec. Wld Hlth Org., 1970, 184, 22 (Resolution WHA23.42)

2 Off. Rec. Wld Hlth Org., 1971, 193, in press (Resolution WHA24.57)

RESEARCH NEEDS

Since drug dependence involves three interacting components, human beings, their socio-cultural environment and dependence-producing drugs, this triad provides a convenient framework for a discussion of research needs. No one of the factors may be ignored in providing services and conducting research, but a particular research undertaking may be focused primarily on one or any combination of the three.

I shall not dwell on the important need for additional research focused primarily on drugs, such as the identification of dependence-producing substances, their mode of action and metabolism, the development of needed effective therapeutic agents having less dependence liability than certain of those now widely used, and the development of suitable, long-acting antagonists for various dependence-producing drugs. This subject is being covered by other essayists. Rather, the substantial need for intensified studies focused primarily on drug-users, their environment and the entire triad will be stressed.

Three areas, in addition to that already mentioned, have been identified as requiring markedly intensified research efforts.[1] The first has to do with the *effects on man and society* of taking drugs at various dose levels for differing lengths of time. Here, we are interested with phenomena that occur between, as well as during, periods of intoxication. Clearly, there should be no particular reason for concern about persons taking a drug capable of producing even substantial psychic or physical dependence if such use were not apt to result in significant public health and related social problems or to produce appreciable adverse effects in a signficant proportion of users. Studies in this area should endeavour to make a distinction between pharmacodynamic effects *per se* and those associated with a society's reactions to a person who indulges in non-normative, unacceptable behaviour. The existence in various parts of the world of varying socio-cultural attitudes towards the use of a given drug will facilitate the design of projects addressed to these problems.

---

1 Cameron, D.C., Drug Dependence: some research issues, *Bull. Wld Hlth Org.*, 1970, *43,* 589-598

The question of the effects of long-term use of different preparations and amounts of opiates and cannabis are cases in point. Are there reliably identifiable syndromes to be seen between periods of intoxication that are associated with the long-term use of these substances? If so, are they primarily pharmacodynamic effects or are they the consequence of participation in socially unacceptable or unlawful behaviour that in itself may restrict access to various avenues of social adjustment and thus lead to the adoption of an unconventional life style? Are the syndromes in any way the consequence of using cannabis?

Without entering into the widespread current debate on the relative merits and hazards of alcoholic beverages and cannabis preparations, it is perhaps worth noting in passing that: both drugs are used recreationally and socially by millions of persons throughout the world, most of whom do not become dependent on them; however, a significant number of users do become dependent, thus creating public health and social problems; and serious problems associated with intoxication from either drug are not limited to persons who exhibit significant evidence of dependence.

Answers to the "effects questions" are essential if appropriate value judgements are to be made about the nature and intensity of efforts that appear to be reasonable in endeavouring to reduce the present and future extent of drug-taking behaviour. Is the actual or potential harm to an individual and to society greater when he takes a drug or when he is imprisoned for simple possession? In relation to what drug? By what route of administration? Length of time in prison? What is the probability that his imprisonment will stop his drug-taking behaviour or significantly deter others? Some of these questions are also related to the need for studies evaluating the effectiveness of various approaches to the prevention and management of drug dependence.

A second area requiring intensified research efforts has to do with the epidemiology of drug-taking. Studies of the prevalence, incidence and natural history of drug-taking will require the use of indirect "indicators" or "indices" in a given locality (e.g. numbers of persons hospitalized for various complications, case registers, numbers of reported thefts of drugs, reports in the literature and many possible others) plus periodic sample surveys of special groups or total populations. The sample surveys will give additional meaning to the index data with respect to the extent of drug self-administration in designated groups or areas.

More information is needed about the personal characteristics of various types of drug-users (e.g. experimenters vs. dependent persons), the immediate and broad socio-cultural forces, and the precipitating factors associated with the initiation, perpetuation and discontinuation of drug use. Not only is it important to learn why certain persons start or discontinue drug use, but also why others possessing essentially similar personality characteristics and exposed to the same socio-cultural environment do not adopt drug-taking as a way of life. Information on these points should be of value in developing increasingly effective preventive and follow-up activities.

The establishment of monitoring programmes in selected high-use areas of the world would facilitate early detection of changes in existing patterns and methods of drug use, and the appearance of new drugs or mixtures on the "drug scene".

The third area deserving of greatly expanded research efforts has to do with evaluating the effectiveness of policies and services for the prevention and control of drug dependence and the treatment, rehabilitation and follow-up of drug-dependent persons. Suitable criteria must be established against which to measure the results of various policies and services. Such criteria should include not only the presence or absence of drug-taking behaviour, but also, among others, its extent if it still continues, occupational performance, and the character of personal and social relationships and activities. Few treatments of any disorder may appropriately be judged by a single criterion. Even persons who die may have been afforded much symptomatic relief during their terminal illnesses.

Beyond the question of criterion measures is the matter of making meaningful comparisons between the effectiveness of different approaches and techniques, including comparisons between treatment and no treatment or placebo.

"Adequate evaluation requires an explicit identification of goals, a classification of cases that will permit the establishment and valid comparison groups, and the use of standardized methods of data collection . . . . Such evaluation cannot be as as precise as many laboratory procedures since full control of the social and psychic variables that may influence the outcome may be impossible. Nevertheless, such work as has been done indicates that valuable data can be secured in spite of these limitations."[1]

Some of the matters requiring further evaluation as to effectiveness are: (1) the relative contribution made by particular medicaments and other treatment measures within a total therapeutic régime, e.g. sensitizing (disulfiram), blocking (naloxone) or substitutive (methadone) agents vs. group therapy and mutual support of the persons involved; (2) the degree to which follow-up services of specified types contribute to eventual therapeutic outcome; (3) the role of compulsion in the treatment, including rehabilitation, of drug-dependent persons; (4) the degree to which preventive educational efforts achieve their purpose with various target groups; and (5) the deterrent effect on existing and potential users of criminal sanctions of various types.

RESEARCH STRATEGIES

Research access to drugs and users

It would be very helpful if those jurisdictions that have not already done so were to establish and implement laws and regulations that would permit accredited investigators (1) to possess drug-materials required for research, and (2) to conduct epidemiological studies without legal hazard to themselves or the drug-users involved.

Collection of comparable data

A WHO Scientific Group recently noted that, without in any way wishing to minimize the importance of endeavouring to develop new and increasingly effective approaches and methods for the study of complex problems, it must be stressed that the results of individual investigations can be related to one another with increasing meaning as they share more common elements in

1 WHO Expert Committee on Drug Dependence, *Wld Hlth Org. Techn.* Rep. Ser., 1970, *460,* 14-15

in approach, method and definition.[1] To this end, WHO is devoting special attention to the development of means to facilitate the collection, collation and dissemination of existing epidemiological and other relevant data along systematic and comparable lines, and to foster needed collaborative studies involving several research centres. The possible mechanisms being considered include (1) the establishment of a collaborative reporting system, (2) the designation of WHO drug dependence research and training centres, (3) the development of a system to monitor selected high-drug-use localities, and (4) the provisions of means whereby persons involved in these and related programmes and representing a variety of disciplines could get together in small groups to consider ways to further needed research and, when indicated, to enhance the comparability of individual studies.

SUMMARY

In this brief paper, attention has been invited to the need for further research on the epidemiology and consequences of drug-taking, the effectiveness of preventive and treatment policies and services and the pharmacodynamics of dependence-producing drugs. The urgent need for more common elements in approach, methods and definitions utilized in research undertaking has been stressed. What is needed is bridges between research projects, not moulds within which to cast them. Some possible mechanisms to help achieve these ends have been mentioned.

[1] WHO Scientific Group on the Use of Cannabis. Wld Hlth Org. Techn. Rep. Ser., 1971 (in press)

DR. FREEDMAN: I am sorry Dr. Cameron that we cannot have a whole conference with your paper as the agenda. I would like to underline before we go ahead the importance not only to see if there is manpower that wish to be trained and recruited into areas because it seems to be really scarce, but the importance of governments and international arrangements being such as to encourage research. I refer to treaties and protocols being written to so that the practical problems of getting research done are squarely faced. We in the United States are beginning to worry that this may not be broadly enough appreciated, and I am happy to hear what you mentioned on this point as well.

DISCUSSION

MR. SHOHAM: Yesterday we had the decision, I believe by our Chairman that in order to achieve real multidisciplinary treatment of the subject we need to do some education, and the educated should be the sociologists. I wish to submit to you however that sociologists might do some of the education. This might not sound modest. A medieval Jewish saint told his students - "Don't be so modest; you're not so great." Those of us who deal with deviant social behaviour have tried to construct a model according to which they were able to study this phenomenon on a real multivariant basis. Again I wish to submit to you that the study of drug dependence is basically this kind of deviant behaviour. We have a certain behaviour of an individual and we also have a human group that has its laws, its 'mores' its beliefs and its mechanism of operation. Now what we do study here is the interrelationship, the interaction, between this given behaviour of the individual and the various normative structures of the group. If this is so any factor which is taken out of its multivariant context, becomes totally meaningless. I recall yesterday the discussion here about the merits or demerits of the economic factors when related to drug dependence. If we take this economic factor out of context it becomes entirely meaningless. In our field, we long ago discarded the notion that economic factors by themselves had a certain direct bearing on our study, because to give you an extreme example, the starving Hindu, having achieved complete economic freedom, would not dream of slaughtering a sacred cow and frying himself a steak. On the other hand if a certain person is unable to buy another mink coat for his mistress, he would define the situation as "needy" enough to embezzle funds from his employer. The definition of the situation is all important and the definition of the situation is always a multivariant affair. So here again if we do wish to make any headway in the investigation of drug dependence, we should look at it basically as an inter-relationship between an individual and a certain normative structure. Multivariant analysis is the only approach possible. In other words, any unit, anybody or any approach towards the investigation of drug dependence should be by definition - multivariant. I agree with Dr. Edwards. But this connectedness is a semantic one, it is not a disconnectedness or a conflict of meanings, either meanings or materials. Well, if the multidisciplinary team starts working and then there is a kind of misunderstanding they might start by sign language but eventually they are going to reach a kind of modus operandi which is based on real understanding of the issue but not a kind of compartmentalization which basically mars the issue and does not lead us to real understanding.

DR. DENICKER: My question is addressed to Dr. Cameron. He indicated that it was possible to measure the development of the phenomenon of drug dependence by indicators or indices. This is just what we have begun to do in France. But he mentioned that it should be possible to distinguish between drug dependents, that is habitual drug users, and occasional drug users. I would like to have some more precision on these two categories, for the more information we obtain from the police, customs, pharmacies and hospitals, the more difficult it appears to us to distinguish between occasional and habitual drug users.

DR. CAMERON: Among the "indicators" that one might think of are (1) the number of admissions to hospital for various medical complications, associated with the use of different dependence-producing drugs, and (2) the number of drug-associated deaths. To give meaning to such indicators, it would be necessary periodically to carry out sample surveys in the locality to determine as accurately as possible the prevalence, incidence and amounts of drug use and the personal and social factors associated with such use. The "indicators" would then be useful guides to the situation between the sample surveys. In considering drug use it is important to distinguish between experimental or occasional users and regular users. It is quite probable that the former groups may have different personality characteristics than those who find it interesting to use regularly. Unless questions of amount, frequency and duration of drug use as well as circumstances of use are taken into account, little value will be obtained from epidemiological studies.

DR. EDWARDS: I would like to give warm support to what Mr. Shoham has just said. I think that the danger with research is that we perseverate in certain approaches and we are so concerned with the days work, the immediate demands, that we seldom have the opportunity to lift our heads imaginatively and look for new and promising lines of endeavour. This is not to blame ourselves. We are under pressures and we respond to pressures by getting on with the immediate and easily perceived job. Social and behavioural sciences have an enormous contribution to make to this work which has not yet been fully explored, and it is useless to ask whose fault this is. One of our greatest dangers at present is constantly looking at the drug addict rather than looking at the system with which he intereacts - if you like, constantly looking at the drug addict and not looking at ourselves. One may wonder what would be the prospectus if in a fantasy world, turned upside down, the drug addict was launching a research programme direct ed at the curiously deviant society which reacts to him so aggress-ively. We are ourselves always asking questions from a particular subculture's point of view. What questions would the drug addict himself see as having priority? I think he would say for instance:

"Let us study the magistrates - let us see why it is one magistrate will send a young cannabis user to prison (which still happens in some countries) while another will impose a small fine or dismiss him with a caution. Is the magistrate's personality the determining factor, or is he simply responding to his particular perception of his peer group pressure, as a man who wishes to conform with his fantasy of what other magistrates think? Is his behaviour determined by basic ethical ideas as to the nature of drug taking?"

Such research as this should, I believe, be more than a fantasy. We could profitably, in any country, be studying the perceptions and the value systems of the courts, and transcultural studies might here also be of value. The young drug user expelled from his school, because he has been found with two amphetamine tablets, might then say to us -

"Let us see why schools are reacting very differently. Why some schools are able to contain the adolescent who has been found using drugs while other schools are so anxious in the face of minimal drug use, that they cannot handle the problem other than destructively?"

Let us then look also at the doctors, and at clinics, which give methadone to some subjects and not to other subjects. Let us try to understand the criteria which are really determining the doctor's differential behaviour in these situations. My central plea here would be to see the practicality of studying the reacting system, whether it be policeman or doctor or, teacher, or even perhaps very daryingly the politician. Finally, I would wonder whether Dr. Cameron would agree to add to his fascinating protocol an item which would acknowledge the fact that we really do need much closer research on how research is used. This may seem prodigal when there is so much research waiting to be done, and yet I believe this to be a key question. If we had the answers to all the questions that Dr. Cameron today proposed, would we have any assurance that those findings would in the real world bear on policies and practical action?

DR. LEONG HON KOON: Allow me to say a few words on some recent experience in the particular problem that Dr. Edwards brought up. About a decade and a half ago I was looking after old opium addicts in a prison institution. And in the course of treating them I often went to their families. I had the job of getting the sons to accept a father who had been imprisoned for an offence which was actually illegal. Before 1941 opium smoking was legal then after World War II it became illegal. A few years ago he was smoking happily, then after that he landed in prison just for doing the same thing. And that practice often separated him from his family and when we tried to rehabilitate him we had to bring back the wife, the sons and the daughters. Now recently, just a few months ago, even a few days before I came here, I had a chance to counsel young persons from the age of 15 to 19. Some of them had dropped out of school, actually sacked, not because they were smoking marihuana but because they were playing truant as a result of taking marihuana. I had to speak to the fathers about their children - their sons and daughters behaviour, running away from home and that sort of thing. It was not an easy task as the reaction from some of the parents was - "Well if he wants to run away, let him run away and let him die in the streets, let him be run over by a car" - something like that. That sort of thing actually could happen, because quite a number of the young persons I saw had taken an overdose and landed in the street and fallen into a drain. One particular person I saw was a bus conductor and his father was running the bus. For some reason he had taken about four MX tablets in addition to his usual doses and then he went in the bus. The bus made a sharp turn and he fell out of the bus and finally landed in hospital. Because I had only one or two weeks before counselled him against taking further pills, I recognized him as a drug-taker.

Another question I am often asked is how big is the problem in any place? It is very difficult to answer that sort of question, because in my own case there are only several places where I meet with the world of the addict. I will say as a prison medical officer in the past, I would meet those people who have landed in prison as a result of violating a dangerous drug ordinance - in the old days it was the older person, now it is the younger person. There is a hardening of the laws against the young drug taker and many a person found taking ganja, which is the local name for marihuana, is often jailed for three months or is fined a sum which he is quite unable to pay. Well fortunately we have a system of being able to transfer these people to an existing opium treatment center. We had actually to convert from the treatment of the old time opium addict to the treatment of the younger offender. That again presented quite a number of difficulties. Then again I often had to counsel schoolmasters. A school child would be brought up by their schoolmaster for having taken drugs for behaving peculiarly in class and they would bring him to one of our patient clinics. Naturally, the doctor there did not know what to do. But since he knew that I had been dealing with this problem he would ring me up. I would tell him that the first thing to do is to take this child into his room and assure her or him that we are not going to make a pronouncement on whether the child is an addict or not. The second thing is to assure the parents that we are not going to make a verdict on the child. Thirdly to ring up the headmistress and say that she is not to state definitely whether the child has or has not been taking drugs. The last and important thing to do is to talk to the child - "Well you have a problem now let us see what it is all about." And that ould probably start off a half hour discussion - a half hour that is very difficult to obtain in present day medical practise.

DR. FREEDMAN: The schools in the United States are under terrible pressure to do something; in fact, everybody is under pressure to act and to show to other people that they are doing something. Their annual reports have to prove they are doing something. In my experience, you can show people what <u>not</u> to do as a first step--how not to panic, how to begin to look at themselves, at their own community--with its own problems and resources--because school systems differ. There are different states of distress if you will. This is a very important useful function that the doctor can perform. But you can also bring up the point of being in the position of a policeman--rather than a counsellor--to an individual--rather than to a system. But that is a whole topic.

DR. EVANG: Thank you, Mr. Chairman. Mr. Shoham and of course Dr. Edwards have brought us by jet up to the high clean air of methodology and science, and now Dr. Leong has brought us down to earth again. May I strike a middle line. Firstly, I would like to congratulate Mr. Shoham on his remarks. What he said, if you translate it into the language that we use in the world of hygienists, would mean, i.e. that you apply the epidemiological-statistical method. Facing multifactorial causation problems, you have to approach your object in a detached and objective way regardless whether you as a scientist run into trouble. I have a little more difficulty in following Edwards. Studies for example of the way in which schools behave are, as I am informed, underway. So are studies of the difference of magistrates in their decisions. I do not think we need to turn the world topsy turvy and ask for individuals with drug problems to organize this type of research. I wonder whether they could. My point in making this is: aren't we by taking this attitude widening the gulf? Are we so different from the abusers? We are all human beings and must of necessity be dependent on something. A person who was not dependent on anything would suffer from or enjoy - as you prefer - a sort of freedom which may be compared to the freedom of a piece of soap which has been lost in the bathtub. My point is this, that instead of widening the gulf, of speaking of two completely different attitudes, we should do our utmost also in the scientific approach to underline what is the common ground for all.

PROFESSOR REXED: Thank you Mr. Chairman. First I would like to thank Dr. Cameron for his exposition of research needs. He evidently had answered by questions of yesterday even before I had put them. I think his discussion on what needs to be done in the area is excellent and one can only hope that it will be possible for WHO to shape a programme along these lines. It is highly necessary that these programmes come into being. We have just decided between some 80 countries, to draw a new convention on international collaboration for the control of psychotropic drugs and I think when this important convention comes into force, the role of WHO will be a very important one. It will not be possible for WHO to play this role unless it institutes quite a different and much larger programme along the lines primarily I would say of the papers of Dr. Cameron and Dr. Chrusciel. I do hope and I think this room is a fitting place since the Executive Board meets here, that it will be possible to get the contributing countries to WHO really to live up to the situation and provide means for an expanded programme. My second remark would be on the question of factors causing the situation which we are discussing. I would absolutely agree with Mr. Shoham and say that his approach is the only one that is possible, seeing the multivariant factors which are at work. I would like to say this is not a contradiction of my remarks of yesterday, which seem to have been understood by many, that I only wanted to point out the economic factors. Of course not. I just wanted to keep the socio economic picture also in the memory of the people who are going to discuss things here. Because even though we and many who discuss these things profess that they do take these things into regards, there are mental inhibitions with all of us, inhibitions that come from our evaluations which play a role when we make our models for research plans and these tend to inhibit us when we come closer to the area which have political implications. Then, we stop and say "this is not research," "it is not science," "it is something else," and we retreat into a more secluded area where we can easier manipulate factors than we can in the broader field where we come in contact with social and political life. I think this should be avoided and if one follows to the bottom the attitude of Mr. Shoham I think you have to take these factors into account and I would be very happy. There is one possibility that would follow in the practical work with people having such problems. If you take the line of Mr. Shoham and I would like to put this as a sort of general question, isn't it rather dangerous seeing the number of factors at work and the difficulty of isolating even causating factors or type of drug from each other? Isn't it rather dangerous to go on labelling these people according to specific diagnosis? We tend now to discuss in this meeting for instance a special type of person that we call a drug addict or somebody who is dependent on drugs and not only that, we separate according to what drugs they are using. We have morphine addicts; we have barbiturate addicts; we have alcohol addicts; etc. etc. I am using the word "addict" just for simplification. So we tend to divide them into drug users and misusers; into alcohol users and misusers; into psychopaths, or whatever that means; into different kinds of psychotic situations, etc. etc. Is this really very important if you look at it from the socio-rehabilitation side? We know from experience that all these situations are often mixed and people switch according to chance from one drug to another. They may use one drug then go on to another, then to alcohol and then they do everything at the same time. The problem behind their behaviour is evidently something else. Some kind of total situation so to speak, psychological, medical and all kinds of things. Perhaps in trying to help these people we should leave these diagnosis alone. Just put out teams of people such as a doctor, a psychologist, a social worker and let them work along the lines of problem shooting or trouble shooting or crisis solving or something like that. They would not be interested in putting any kind of label on the man or woman they are helping but rather try to look at the situation as a whole and interact in all the difficult areas. We would have to reorganize much of our social and medical services and much or our thinking too, but I wonder if we do not have to go along such lines if we are to get to the bottom of these problems.

DR. FREEDMAN: I have noticed a great deal of difference in this conference and the few I have attended a couple of years ago in this great country. I have heard the word "addiction" and "addict" usedfar, far less, and a greater recognition of variety, of patterns of people and the significances of the fact that drugs happen often to be involved with behavioural styles, and I think these developments represent an authentic move ahead.

DR. EDWARDS: I would like to try briefly to answer Dr. Evang's comments, by saying that I think that he and I share common ground with everyone in this room as to belief in the common nature of man, despite man's many differences. Where I would not share common ground with Dr. Evang - perhaps he would respond to this and put me right - is that I do not believe anything that I have said is likely to set up greater differences or to be divisive or schismatic. I believe on the contrary that it is blindness to suppose that we are not at this present moment already being highly schismatic, and I would invite Dr. Evang to look at himself, myself and at our distinguished colleagues, sitting around this table, considering the drug problem. I am not very good at guessing ages. I should not think that anyone here is aged under 30 and would have thought that the average age was considerably older. I would not think that there were many cannabis users sitting around this table - I may be very wrong indeed. I would not have thought there was anyone here whose confessed credential was that he was a former heroin user. I am uncertain as to whether opium users, cannabis users, or a man whose credential was as ex-heroin user would necessarily be good research workers. I think, however, that they would have been enormously useful to us in identifying the relevant problems. I believe that our present institutions are set up violently to confirm the difference between the person whom we label as a drug user than the person whom we label as expert. And I think rather than showing a slide to make that point, we simply have to show this room. It speaks that message so loudly and so obviously, that we must choose normally not to hear. As to the question of what research is already going on along the lines that Mr. Shoham and I would propose, on the reaction of social systems to the drug using individual - of course there is some work going on, but I do not think there is enough. Dr. Evang's knowledge of the literature may well be able to catch me out on this but I do not myself, for instance, know of any published study which relates the value system or the knowledge of teachers to the way they react to school children who take drugs, but I wait for my ignorance to be repaired.

DR. EVANG: Thank you very much, Mr. Chairman. Of course, I should be the last to respond, due to my age. However, I have the privilege in spite of my age to be in constant touch with these youngsters, and we have no difficulty to have quite an open dialogue. I am often asked this question: "How can you have an opinion on cannabis if you have not used it? - how can you have an opinion on heroin who hasn't used it?" My answer is, "I cannot form a sound opinion on the experience which you, and you, and you - had with cannabis. I need the experience of hundreds - thousands." The effect of the drug differs so much from individual to individual, mainly due to his social and mental situation. Therefore if I was going to judge from my own experience, alone, I would most certainly be mislead. Sometimes I finalize by saying "Have you ever heard of a man that gave birth to a child?" When the "No" has been said: "And you take the attitude that a male obstetrician is out of the world?" It is still surprisingly difficult to make laymen accept that the scientific attitude generally asks for big numbers and that also in this case we can only use that.

DR. FREEDMAN: We have long experience in our rehabilitation programmes in working with ex-addicts, and it is terribly useful. But we have found, Dr. Edwards, that the cautions you speak of are terribly important. These persons end up by being quite human, but limited in vision and invested; yet, at the same time, you learn so much if you consult them, if you let them--to an extent--guide you. The problem is collaboration in research. Some day, because of the history of this era, somebody ought to very seriously --without any person attack on the man and personalities involved--examine the evolution of Tim Leary's thinking on this; he moved from what I thought was very sound social psychology, with a notion of collaboration between subject and experimenter and an understanding of how the exerperimenter often predetermined the results, to the position of making the experiment himself. In losing focus on the problem, in tipping the balance too far, he lost the power of the scientific experiment.

DR. DUBE: I am trying to think about the problem from a common sense point of view. Here we have a habit which has been quite extensive almost endemic in the society. In the past we have never bothered about these people who indulged in the habit. Now that the students have started using it and the young people have started using it we have suddenly become very much concerned. We ascribe all kinds of causes and describe it as student protest, psychopathic tendencies, some weakness in personality, etc. I wonder if it is not a kind of general phenomenon of changing customs, changing values and changing fashions, like fashions in hair style, fashion in clothes. Probably such a thing is happening. I think that the student population in my country is being affected by such influences. I also ask if ordinary reformist methods may be of value and if so why it should not be utilized to tackle the drug problem. To give an example: In my country, and probably this may be the experience in some other countries too, that in certain classes there was wide use of alcohol and other drugs. Suddenly due to the changing political and social scene in independent India and the influence of Ghandi, many zealous social reformers influenced their social groups and advocated that "this drug be banned". Suddenly the whole society completely stopped using it. The efficiency of this simple technique should be investigated, I would like to ask if the government propaganda machinery is good enough to help stop the habit.

PROFESSOR BASH: Thank you Mr. Chairman. I really wanted to reply to Professor Rexed, but I first want to say that I agree with Dr. Freedman in his great satisfaction that the word "addict" has not been used so freely and generously in this meeting as sometimes happens. Professor Rexed, if I understand him correctly, would rather dispense with psychiatric diagnosis or put it in a secondary place. I feel that we are running a danger of rejecting important and useful information. Certainly psychiatric diagnoses do not have the degree of surety as physical measurements, but they have a certain usefulness none the less in categorizing and sometimes comprehending. They also have a practical value. We must remember that while the problem of drug misuse or dependency is large, our means of treating it or dealing with it in any fashion are limited. Speaking now not as an epidemiologist but as a practical clinical psychiatrist, we have to establish priorities of treatment. The priorities we choose depend to a certain extent, I believe, on the diagnosis we make. We feel more inclined or regard it as probably more rewarding to spend considerable time on a neurotic user than on a confirmed moronic, possibly already brain damaged psychopath who is much less likely to respond to treatment. His diagnosis must be related to the social context and background and to the pattern of social change, but I would not like to see it thrown out altogether.

DR. FREEDMAN: Thank you. I would note a movement in social medicine in the States (where self-help groups are doing a lot of the treating) we could call "Operation Outreach"; this is very, very important and a great comment on the shift of the medical profession, which really has been at a great distance from its constituency. The problem that is beginning to come to the fore in the free clinic kind of operation is the rebirth of professionalism. Milton Friedman of the University of Chicago (known for a certain kind of Economics) made me think of a fantasy. If all drugs were freely available, anyone could get insulin in order to feel dizzy any time he wished--absolutely no laws. Pretty soon you would have consumers unions, special drug stores and pharmacies that the elite would recognize as being sound. You would end up with manufacturers who were believed and those who were not, and with some sort of certified person that had some kind of special knowledge to dispense drugs sooner or later. I think this also applies to diagnosis where it comes into the continuum of health concerns.

PROFESSOR KIELHOLZ: I should like to ask a question to Dr. Cameron. What is the programme of the World Health Organization in relation to the coordination of research such as coordination with the Council of Europe and coordination with research centres on the national and international levels? We need to have the possibility to coordinate research and exchange experiences in research and collaborative studies.

DR. CAMERON: I am not sure that I can fully answer your question Professor. One thing that we hope to do soon is to convene a small groups of experts to try to help devise guidelines for recording and reporting information that is already being collected by various institutions and workers so that it would be more comparable than it now is. This involves, e.g., agreement upon definitions, age groupings and the use of cohort data. A second approach would be to enlist the aid of certain institutions and individuals that would collaborate in collecting and reporting certain types of information. A third activity would involve designation of the research and training centres mentioned earlier. From among the persons involved in the data gathering system and the research and training centres, and perhaps other scientists, there could be established a cadre of experts from which smaller groups would be brought together to deal with particular questions. Among the questions might be (1) the kinds of additional research urgently needed, and (2) the means by which bridges could be built between studies - bridges involving approaches, methods and instruments to be used. Other small groups, with some overlapping membership, might be convened to address themselves to other questions. When these clinical and research personnel get together there would be a great deal of interchange of an informal nature as well as of a formal nature. We visualize WHO as providing mechanisms that will enable people who can contribute to the field of drug dependence to meet and exchange ideas around a variety of important themes. We are not so concerned with conducting research ourselves as we are in fostering comparability between studies that are carried out by others and furthering the development of research of a cross cultural nature.

As you know, Professor Kielholz, the WHO Regional Office for Europe did work with the Council of Europe in gathering data about the existing patterns and the extent of drug use in a number of European countries. This was a beginning and you yourself were a key figure in that undertaking. We have worked with the Council of Europe and other organizations on other projects and have endeavoured to help some of them when they organized meetings relative to dependence on alcohol or other drugs.

DR. FREEDMAN: I think the one topic that we all should be thinking about is why it is difficult to crystallize research programs and support. For, after five years of crisis in our country, nobody can get off the stick and get research mounted in any kind of way that would be useful and focused. Why it is that there is such swirling around and very little locking into gear in this area I don't know. It may be the intrinsic universality of the drug problem that is difficult to focus. This kind of thing you speak of seems like a useful device through which those who give us money and organize in this way might finally be able to respond. I think it is a need.

MR. STEINIG: This has been a stimulating debate on a problem facing the International Narcotics Control Board.

My experience in the field of the control of narcotic drugs is limited to the international legal and administrative efforts of the society of sovereign States and the results of these efforts. You are, of course, aware that international relations are predominently relations of States. The principle of absolute national sovereignty, however, cannot be nowadays openly invoked and acted upon, even by its most determined defenders, with the uncompromising rigidity normally observed during the period before 1914. The development, since 1925, of international legislation dealing with the control of narcotic drugs illustrates well this evolution.

Dr. Edwards' and Mr. Shoham's remarks brought back memories of the successive stages of this evolution in the field of narcotics control, and the accompanying difficulties, during the last 50 years, In the beginning, under the League of Nations, the legal basis of the international control of narcotic drugs consisted of one international treaty only: the Hague Convention of 1912 which, as I mentioned, did not set up any organ of international control. At present there are 10 international treaties regulating the production and use of narcotic substances; the 11th Treaty, the Convention on Psychotropic Substances, concluded in Vienna on 21 February 1971, is now in the process of ratification.

Our exchange of views so far suggests to me some questions. Is not the worldwide spread and the scope of drug abuses due i.e. to the erosion and, in many instances, to the complete destruction of the authority of and respect for basic human rights which were the essential elements in building the 19th century's feelings of stability, security and protection? Is not the present lack of these feelings seriously aggravated by the fact that the faculty of human adaptability, even of very young people, has its limits, that in the present situation this faculty is unable to cope successfully with the unprecedented acceleration of the pace of radical changes of the conditions of human existence the world over and with the resulting all pervading uncertainty about the future? Should we not study also these phenomena when we are exploring the causes of the manifest massive flight, particularly of the young people, away from deeply disquieting social reality, their evasion of this reality by withdrawing, with the help of drugs, into the delusions and hallucinations of artificial paradises?

How does society react to these fugitives? In an alloplastic civilisation, society is constantly, not to say compulsively, bent upon changing its environment, with the help of rapidly expanding modern technology, in order to adapt it to its concept of progress and "happiness for the greatest number." In this type of society the reaction to the mentioned tendencies of evading social reality is rather severe: it assimilates sometimes drug dependence to a crime and opposes it with moral and legal repressive measures.

In an autoplastic civilisation, society is guided by the conviction that the natural order and balance of things must not be disturbed, that changes of man's environment should not go beyond the necessities of sustaining life and that it is rather man who should adapt himself to his natural environment. Such a society is, on the whole, more inclined to tolerate the evasions of social realities than a society of the alloplastic type.

It can be assumed, that one of the causes of the phenomenon of young people wandering from country to country, from continent to continent, is a desire to escape from the rigours of a civilisation pursuing continuous change and find, in an autoplastic society, tolerance for their evasion of social reality and their withdrawals into artificial paradises.

A last remark about the need of distinguishing sharply between the motives of a drug-dependent person and of a coldly calculating drug trafficker motivated by predatory greed. In this gathering it is superfluous to dwell on this difference. I am mentioning it only because it took a long time to dispell the erroneous view, still held in some quarters, that legal-administrative measures of the prohibitive and punitive type along would eliminate both drug dependence (including the demand created by drug dependent persons) and the illicit drug traffic as well. According to this view drug abuse being considered a crime, the drug dependent person was treated as a criminal. As the INCB pointed out (in its report for 1968), the differences between the motives of a drug dependent person and those of the trafficker call for different means and methods of restraining their activities. Now there is in many countries a change in the approach to the problem: for the punitive method of dealing with drug addicts is being substituted their medico-social treatment requiring specialized techniques adapted to different types and causes of dependence. It is now rather generally recognized that the whole problem calls for continuous study, extension of research facilities and trained personnel.

PROFESSOR OLE WASZ-HOCKERT, M.D. (International Pediatric Association): The IPA wants to congratulate the organisers for this "door opening" meeting. The pediatricians are involved in this problem. After all, it is our patients who we now discuss. Pediatricians are in a key position to do part of the basic preventive work: the close contact with the mother, the parents, the school and above all, with the child and later the adolescent.

The experimentators, those youngsters who try cannabis, a couple of times are influenced by a new culture, a new fashion, but the real problem is caused by the users. And now we ask ourselves what makes the user? Here, in addition to those research projects suggested by Dr. Cameron, I propose that in different parts of the world a number of research-groups should be established including pediatricians, child and adult psychiatrists, psychologists and especially - as Dr. Shoham remarked - sociologists, to study on a casework basis the family dynamics of the users. The early mother-child relationship, the parent attitude and upbringing, the school environment should thoroughly be studied and analyzed in order to find out the "risk-group families" so that preventive work can be effectively directed to them. I suggest that the CIOMS and the WHO organize a workshop to plan research and that the pediatric approach should be taken into consideration.

DR. CAMERON: Mr. Chairman I think it might be useful if I were to comment a bit further on the question of monitoring high drug-use areas earlier mentioned, and also the question of convening small groups to work on particular problems. One might think of drawing together persons who are already working at the street level with drug users in high drug-use areas, e.g. persons like Roger Smith, sitting here, who are providing services to young people in difficulty, including difficulties with drugs. Comparable persons are to be found in Uppsala, Sweden, Soho in London, the Kings Cross area of Sydney, Australia and Singapore among others. There are a number of such persons around the world who are already deeply engrossed in this work. They are, or could be, collecting and recording important data. It would be useful to bring these people together and ask them to tell each other what they are doing: They might be asked if it would be possible to agree on some common definitions so that comparisons could be made between their data. What additional data would they like to collect? What additional help would they have to have if they were to collect these additional data? Would it be useful if a small grant could be obtained to assist in data collecting? We cannot tell people what research to do but we may be able to take advantage of their existing interests, provide a mechanism for cross fertilization of ideas, and stimulate collaborative endeavours. This is but one example of the many possibilities.

DR. MENDELSON: Very briefly, I would certainly concur with Dr. Cameron, returning to his discussion. I think it would be a very worthwhile enterprise to try and establish some uniform data collection, processing and evaluating system. It would be great if we had that between two or three cities in the United States or be it on transcultural, or a transcontinental basis. Let me just make one very brief remark - it is a feeling of caution that I have, that I will share with you. It has to do with setting research priorities in terms of gathering a group of individuals who presumably have interests and knowledge "expertise" decide upon which areas should be focused upon in investigative work. This is a good idea, but often I feel it can mislead as well as serve a good purpose. Setting research priorities is an extremely difficult thing. Now for those of us who have some feeling that - while we like to be eclectic, catholic, non-parochial, look at a wide variety of various "set no hierarchys" in terms of the potential contribution that any area can make towards solving the problem, we all do have our special skills and our special interests and when it comes to funding our special needs.

So it seems to me setting research priorities perhaps can be done on a basis that would be a little more wieldy and perhaps not carry so much weight - an international body did this - unfortunately it might influence our political colleagues who provide the funds to do the work to really think that we do know what should be done. We think we know, but we are not sure.

The other thing is the involvement of what might be called the consumer or any special constituent, be it on the basis of criteria of age - special experience with the drug problem and so forth. These individuals can be extremely helpful. Perhaps I am not going to be as kind as Dr. Freedman who says "they are really great people" - yes and some of them are, but I think what we really should do is make a good attempt to assess the efficacy of the contribution that these groups or individuals can make, not simply because it is nice; of course we want to be encompassing on an A priority basis say, that because an individual has an experience with the drug he might be uniquely somehow better able to help other people who have had that experience. I think we have to ask self groups to really show us that they are making a contribution and not merely accept their contribution on the basis of good will.

DR. R. ELLING (Chief, Behavioural Sciences, World Health Organization, Geneva): Mr. President, Mr. Chairman. I wanted to indicate my excitement at that part of the discussion dealing with the question of society and its problems. I want to make the suggestion that we ask ourselves seriously where the problem is resident. Is it in the addict? Or is it in the structure of the society, the group formations of which the person is a part? I am reminded of some work of Professor Robert Merton on Social Structure and Anomie*- where the individual who is exposed to a set of expectations in the society and then has no means that are really reasonable or adequate to meet those expectations is bound to experience a high level of frustration and at that point seek some way out, drugs being only one of many. I think some very serious work as suggested by others this morning could be done on societal purpose and organization, on group formation, and on purpose and ability of individuals to fulfil purpose within a group. I don't know how proper it is to point to the work in this context, at this time, but I was very fascinated reading a book by Jan Myrdal** who examined the same village in China at two different points in time. He found that some serious change had taken place whereby, among other things, there was an excitement on a part of the people as to what they were doing and they were involved with a sense of common purpose - and the problem of drugs had disappeared. In quite a different context, and with more limited implications, there is the study of teen-age gangs in New Jersey, including some which went off hard drugs. The individual is seen as responding to the approbation of his group of significant others and the group as a whole is seen as exposed to an environment of constraints and opportunities.*** Some more fundamental work along these lines, I think, would be highly desirable.

---

* R. Merton, "Social Structure and Anomie" and "Continuities", pp 131-194 in Social Theory and Social Structure, revised and enlarged edition, Alencoe, Illinois: The Free Press, 1957

** Jan Myrdal, The Revolution Continued

*** Julius Klein and Derek L. Phillips, "From Hard to Soft Drugs: Temporal and Substantive Changes in Drug Usage among Gangs in a Working-Class Community," Journal of Health and Social Behavior, 9 (June 1968), 139-146.

DR. ANDREASSON: My question belonged to the discussion between Dr. Edwards and Dr. Evang. This is perhaps overplayed by now, but we are still discussing research. Very briefly I would like to inform you that I took part in an international meeting some years ago, where misusers of alcohol and drugs were invited to report, and the objectives of the meeting were just to get new guidelines for research. This meeting ended in laughter, fun, noise and I can say it ended in chaos. Nothing came out of it. Now my question to Dr. Edwards. Have you any experience of a serious Round Table discussion with misusers of alcohol and drugs, just in order to find a ground for new lines in research?

DR. EDWARDS: I think that perhaps the mere size of this table - the distances between people - creates by accident polar positions where no polarity really exists. I am not for a minute proposing that age debars a person from making a contribution, and I am sure Dr. Evang will make an even finer contribution at the age of 80. What I am suggesting is that neither does youth debar people from making a contribution. Male sex does not mean that you are unable to make a useful comment among obstetrics, but at the same time I do not believe that the fact that you are a woman should disqualify you from engaging in such debate. I would indeed be very sad to see any obstetric forum without a woman present. I don't really regard it as particularly revolutionary to suggest that a person chosen not just because he has experience of drug use but chosen also on the grounds that he is a serious person and competent in discussion, should join our sort of discussion. But he should not be allowed to dominate this meeting or to replace Dr. Evang, but rather to supplement Dr. Evang's wisdom. Coming to your question as to whether I have experience of consulting alcoholics or drug users in the process of designing research, the answer is "yes". To do so is not particularly revoluntary but a basic and common-sense technique of enquiry and one to be applied whenever you are tying to identify the nature of an important problem out there in the community. When you ask people who are supposedly afflicted by a problem to engage with you in trying to identify the nature of that problem, they will then often tell you that the problem - in the words of the last speaker - really doesn't reside particularly in them. Again, this isn't to be seen as threatening - it is not a wicked thing to say that the policeman, the doctor, all of us, are legitimate subjects for study. I think we should today be rather beyond seeing that sort of remark as in any way contentious. We have moved into an era when we can without undue anxiety, accept the fact that the drug question involves a study of society as well as of the individual.

DR. FREEDMAN: The session has been stimulating, so stimulating that it is impossible to summarize. Practically everything has been mentioned. I have one further topic that I think at least should go on the record--the study of prescription drugs that are being abused. There is a recent one completed in the United States. It shows, in the last ten years, barbiturates and amphetamines decreased relatively as prescription drugs over a long trend, while there is an increase in sedative, anti-anxiety drugs; but doctors are not apparently writing prescriptions for a long period of time; only 8% of our population are on the sedative, anti-anxiety drugs or stimulants for longer than six months. It is a much less panic-producing picture of prescribing behavior than I had anticipated. This has to be matched against some of the concern about sedative, anti-anxiety drugs and their specific harm. In my reading of the great London gin epidemic of the 18th Century, which Hogarth so beautifully portrayed with the various different social classes and their alcohol problems, it took a wide variety of measures to stop that epidemic, including economic measures, taxation, clever taxation--making it impossible for the wholesaler of alcohol to sell on credit, which is a profound piece of behavioural control. It took the clever disposition of taverns, it took alternatives such as coffee shops. It also significantly took something like a cause, something like the right war. It took John Wesley (and evangelical Methodism of all things) approving, as well, the alternative of beer. With all of these together, the consumption of gin dropped within 20 years something like from 11 to 14 million gallons to 1 million gallons. There might be some hope, and if there is a real John Wesley in this room, I hope he will stand up.

FOURTH SESSION

PREVENTIVE APPROACHES TO DRUG DEPENDENCE

MODERATOR: Dr. K. Evang

DR. EVANG: The Fourth Session is open. We will now cover the 4th part of the agenda, under the heading of Preventive Approaches to Drug Dependence. We shall start with a contribution from Dr. de Moerloose.

DR. de MOERLOOSE * : Mr. Chairman, when we were requested only a few weeks ago to prepare a paper on the principles of and trends in the legislation of different countries in the world concerning the misuse of dependence-producing drugs, you will understand that it was not without some hesitation that we accepted the task. We had all the more grounds for hesitating, since we were accorded little more than three weeks in which to accomplish the task.

In contrast, in the Seventeenth Report of the WHO Expert Committee on Drug Dependence, published in 1970 as No. 437 in the Technical Report Series, it is mentioned that a Sub-Committee on Penal Aspects of Narcotic and Drug Dependence of the Council of Europe was undertaking "a three-year study in European countries of their laws on the subject, the effectiveness of these laws, and the assumptions underlying them."

---

* The full text of the study referred to in this intervention is found in Annex I, page 254.

Consequently, the paper which has been presented to you is subject to modification and any critical observations, notably as regards the description of the legislation of your own countries, will be most welcome and we would greatly appreciate any remarks you have to offer.

As you will note on examining the bibliography to the paper, two basic sources were consulted in its preparation. In the first place, mention should be made of United Nations documents under the heading "Laws and regulations promulgated to give effect to the provisions of the International Treaties on Narcotic Drugs," published in the United Nations E/NL series; these constitute essential working material for anyone wishing to analyse the legislation of the different countries within the context of the implementation of the international conventions. It should moreover be noted that various countries also refer in their legislation to the recommendations of WHO expert committees, as for example when amendments are made for the purpose of incorporating into the legislation new substances liable to produce dependence. While the information thus published by the United Nations is certainly most valuable, it fails of course to give a complete picture of the national legislation of particular countries since the material must be interpreted in the light of the existing pharmaceutical legislation. Moreover, no systematic information is communicated to the United Nations on a large number of substances liable to produce dependence but not included in the various Schedules of the Single Convention. This gap is filled by the WHO *International Digest of Health Legislation* which regularly publishes the pharmaceutical legislation of the different countries. The treatment of victims of drug abuse is, in addition, an important aspect of national legislation rarely included in the information submitted to the United Nations. This gap too is filled by the *Digest*, which regularly publishes legislation on the treatment of drug addicts; it was essentially this source which enabled the publication, in 1962, of an initial comparative survey on the subject.

Before dealing with the actual subject-matter of the study, a cautionary remark is called for at this juncture, namely that the study is not in any way designed to formulate an assessment or critical analysis of the legislation of individual countries. While the author may have put forward certain personal opinions as regards the legislation, they should be interpreted as such and not as an expression of WHO policy, In this context however we might quote Dr. Edwards: "It would be fascinating to examine the historical evidence that can be culled for the success or failure of legislation." This is certainly an area full of unknowns.

What do we find if we examine the legislation, consisting as it now does of hundreds of texts? The first observation to be made is that in broad terms the legislation deals with three major areas, viz. the licit market in drugs, illicit trafficking and, finally, the treatment of drug addicts. As far as the licit market is concerned, most countries have laid down provisions to prevent the diversion to deviant uses of lawfully circulating substances liable to produce dependence (and intended for medical or scientific purposes). The illicit use and traffic of these substances are dealt with by provisions incorporated in penal law or also in pharmaceutical or special legislation.

An examination of the literature cannot fail to reveal the fact that present legislation is regularly subjected to sharp criticisms. Nevertheless, however imperfect the legislation may be, one may well ask what would have happened if there had been no such legislation. It must moreover be recognized that the legislator may be in a particularly difficult situation. The legislation must of course be constantly brought up to date in order to take account not only of the obligation to adapt its provisions to the international conventions or WHO recommendations but also of the unceasing invasion of the market by new substances whose classification is becoming more and more difficult. An examination of the schedules or annexes to the pharmaceutical legislation of certain countries is sufficient to convince one of the difficulty of this task. Moreover, the legislator cannot ignore the frequently divergent attitudes to the subject, as for example the demand that cannabis be legalized, put forward in recent years by diverse, often influential, groups. In certain countries, it is clear that the legislator, in spite of his desire to eradicate drug abuse, is subjected to economic or social pressures or must take account of the attitude of the society in regard to certain customs or habits. It may not be inappropriate to mention here that some indulgence should certainly be shown towards the physician, pharmacist and even the judge charged with the task of enforcing the provisions of the legislation. The latter is particularly complex for instance in countries with a federal structure where account must be taken not only of the federal legislation but also of that of the different jurisdictions making up the federal

State.

In the paper, emphasis has been placed on the importance of two types of legislation, viz. the French legislation and that of the United Kingdom. These two types of legislation have exerted an influence on the legislation of a very large number of countries. In France, the various poisons used in medicine, including the so-called classical narcotics, have for many years been incorporated in a single decree. The diverse provisions dealing with prescriptions, the refilling of prescriptions, counterfoil books, etc., vary according to the Schedule in which the particular substance appears, there being three Schedules in the decree. As is the case in other countries, the legislation also provides for the exemption of a certain number of pharmaceutical preparations containing substances which may be habit-forming, this being dependent on the nature, composition and especially the concentration of active principle, and we heard from Dr. Mabileau that excepted preparations, as was the case with elixir of paregoric, can be misused. In the United Kingdom, narcotics have been covered by the Dangerous Drugs Acts while other poisons, including some substances liable to produce dependence, have been regulated by the Pharmacy and Poisons Act and the Poisons Rules. Texts of this nature are encountered in many English-speaking countries, the former category being entitled "Narcotic Drugs Act" in some cases. In 1971 (27 May) under the Misuse of Drugs Act the "controlled substances" are now incorporated in the Act under three schedules, as was the case in the United States of America in 1970 (Controlled Substances Act, which contains five schedules).

The paper also gives a description of another type of legislation, i.e. that of Norway, for which we possess very complete information. There are several points of interest, as for example the fact that as early as 1965 amphetamines were included in the Regulations concerning narcotic drugs (issued on 6 January of that year), the most important provisions of the Regulations being applicable to amphetamines. Moreover, while the Law of 20 June 1964 does not automatically oblige members of the medical and pharmaceutical professions to notify such cases of drug addiction, practitioners of these professions, as well as the police and temperance boards, must, if requested, inform the Director-General of Health Services of confirmed or suspected cases of drug abuse. The Director-General in turn supplies these various categories of persons with the information he possesses in this field. Insofar as hallucinogens have not been registered as medicaments, their sale is not permitted. As to hallucinogens, it is of course evident that other countries have introduced similar provisions. Of particular interest in Norway are the conditions governing the prescribing of substances liable to produce dependence; the Director-General of Health Services has issued special instructions on this subject, imposing differing requirements for prescriptions for four categories of persons.

It can be stated that in recent years countries have introduced more stringent legislative provisions, either in the form of amendments to the Penal Code or by the introduction, as was recently the case in France and the United Kingdom, of special laws on drug abuse. Dr. Mabileau has described the French Law. A clear distinction is however made in these texts between the addict who contravenes the legislation and the criminal who makes his living from drug trafficking. The importance of this distinction was stressed by Mr. Steinig. Severe penalties are generally prescribed for incitement of minors to drug addiction. Finally, the provisions governing physicians and pharmacists guilty of manifest abuse in the drawing up of prescriptions or the supply of drugs are becoming increasingly stringent. The penalties imposed on offenders may even include temporary withdrawal of the right to practise their profession. The legislative provisions on drug addicts, formerly embodied in specific texts, are now tending to be incorporated in more general enactments which also deal with alcoholics and even, in some cases, mental patients. There is a manifest trend towards the compulsory notification of drug addicts. Various criticisms have been made on this subject and it is not without interest in this context to point out that the legislation in some countries is more flexible, as for example in Switzerland where the Federal Law on narcotics authorizes physicians and pharmacists to notify cases of addiction of which they may become aware in the practice of their profession. Most legislative texts provide for the treatment of addicts, on either a voluntary or involuntary basis, the decision in the latter case being either administrative or judicial. However, even in cases of voluntary treatment, with its well-known risks and dubious benefit, the legislation provides that the patient may under certain conditions be detained in the centre for treatment.

In general terms, it may be affirmed that the study of national legislation is of very considerable interest, since there is no doubt that international conventions can be drawn up solely on the basis of the prevalent national legislative measures. Moreover, studying national legislation facilitates an understanding of the gaps in the legislation in certain countries, and this may enable the necessary amendments to be made. Another pertinent general comment is that the task of the physician vis-à-vis the drug addict may be particularly difficult, since he is often called upon to act as judge, as in cases of commitment of addicts or alcoholics. The judge's task too may be extremely complicated, since he is called upon to pronounce judgements in matters where the medical factor is preponderant. Finally, it must be recognized that the difficulties confronting both the legislator and the public health services responsible for implementing the legislation have become very substantial and that, without the establishment of special organizations within the health ministry, the implementation of the legislation is impossible.

In the control of drug misuse, legislation, is of course only one of the means, and we may perhaps again quote in this context Dr. Edwards: "How often has legislation blown bridges?"

DR. EVANG: Thank you Dr. de Moerloose, we are certainly very much impressed with the valuable work which you and your collaborators have been able to do at such short notice. Our next speaker is Mr. Steinig. Mr. Husain will then speak on the educational aspects.

THE INTERNATIONAL APPROACHES TO DRUG CONTROL

Leon Steinig

Member of the International Narcotics Control Board

The origins of the use of narcotic substances for medicinal purposes are lost in the mists of the past; so are also their uses for satisfying man's propensity to evade reality by retreating into the illusions of artificial paradises. Sixteenth Century B.C. sources provide the earliest references to the use of opium medicinal ends.*

Efforts to deal internationally with the use and misuse of narcotic substances are of a very recent date. They started in the Twentieth Century with the meeting of the International Opium Commission (IOC) which convened at Shanghai in 1909, largely at the initiative of President Th. Roosevelt. From that first international meeting of representatives of thirteen governments**

* As mentioned by Terry and Pellens in "The Opium Problem" (New York, 1928), pp. 53-55, according to Fielding H. Garrison "An Introduction to the History of Medicine" (Philadelphia, 1922) the Ebers Papyri (their assumed date is about 1550 B.C.) enumerate opium among other remedies. A later source, the 7th Century B.C. "Assyrian Medical Tablets" ("The Assyrian Herbal"), in Dr. R. Campbell Thompson's translation, include opium in a list of 115 vegetable drugs.

** Austria-Hungary, China, France, Germany, Great Britain, Italy, Japan, the Netherlands, Persia, Portugal, Russia, Siam and the United States of America.

to deal with the problem of opium, "its alkaloids, derivatives and preparations," dates the recognition of the abuse of narcotic drugs as an international social problem.

It would be of interest to explore why this reaction to the insidious danger of using indiscriminately narcotic substances came so late, and why, once started, the progress of international legislation in this field was relatively rapid. But time urges to return to the IOC and to mention very briefly the results and consequences of its work.

The nine resolutions unanimously adopted by the Commission and the Records of its meetings indicate clearly the view of its members that the problem of drug abuses required effective international cooperation; that its solution by national efforts alone could not succeed because it transcended the capacity of any single government operating in isolation. The foundations for the subsequent international efforts to stem the spread of drug abuses had thus been laid in 1909.

In 1912 the First Opium Conference met at The Hague. Attended by diplomatic plenipotentiaries empowered to sign an international treaty, the Conference adopted The International Opium Convention of 1912. That Convention transformed the moral obligations of governments, resulting from the resolutions of the Shanghai meeting concerning effective international cooperation, into legal obligations under international treaty law. But no international organ had been established to assist in and supervise the implementation of that Treaty.

Parties to the 1912 Convention agreed:

(i) to limit to medical and scientific needs the manufacture of and trade in medicinal opium, morphine, heroin and cocaine and their respective salts, and every other alkaloid of opium "liable to similar abuse and productive of like ill-effects." Each Party has been left free to determine the quantity of its medical and scientific requirements;

(ii) to control the production and distribution of raw opium, but no control measures were envisaged over production and distribution of opium and no direct limitation was placed on the quantity to be produced or distributed;

(iii) to accept, but only in principle, the suppression of, trade in, and use of opium prepared for smoking;

(iv) to establish a system of governmental licensing to control the manufacture of and trade in drugs covered by the Convention;

(v) to exchange, through the intermediary of the Ministry of Foreign Affairs in the Netherlands, detailed statistical information on the trade in the drugs specified in the Convention, as well as the texts of laws and regulations promulgated respecting matters dealt with in the Convention. No international organ had been established to assist in the implementation of the 1912 Convention.

The IOC and the 1912 Convention have released a chain of events and a sequence of experiences which have gradually disclosed the true nature of the problem, the unforeseen difficulties and obstacles which had and still have to be surmounted in order to build a sound and solid protection against the abuse of dangerous drugs, and its social consequences.

Efforts to bring the 1912 Convention into effect were interrupted by the First World War. The Peace Treaties brought it automatically into force amongst their signatories. Moreover, a special provision in the Covenant of the League of Nations (Article 23 (c)) entrusted it "with the general supervision over the execution of agreements with regard to . . . the traffic in opium and other dangerous drugs." This was, half a century ago, a solemn and legally binding expression of the general opinion that, breaking across the national frontiers, the problem of the abuse of narcotic substances was not confined to any one continent or civilization, that a sustained universal effort was required to deal with it and oppose successfully the internationally operating networks of traffickers who, prompted by predatory greed, were amassing fortunes by exploiting the morbid craving of drug-dependent persons.

In 1921, the League of Nations' Advisory Committee on the Traffic in Opium and Other Dangerous Drugs, set up under the League's Assembly resolution of 15 December 1920, began its work; its first task was to make the 1912 Convention effective in application. Analysing information furnished at its request by governments, the Advisory Committee concluded that the amount of narcotic drugs manufactured by duly authorized factories have considerably exceeded legitimate (i.e. medical and scientific) world needs. The disclosure of these facts, revealing the weaknesses of the 1912 Convention, prepared the ground for the Second Geneva Opium Conference (November 1924 - February 1925) which, according to its terms of reference, should have elaborated a Treaty limiting to legitimate needs the manufacture of narcotic drugs and the production of narcotic raw materials (opium and coca leaves). When, after long and searching debates, it became evident that the Conference will not adopt as a basis for the new treaty a system of quantitative limitation of the production and manufacture in question, the American Delegation, followed soon by the Chinese Delegation, withdrew from the Conference. The Conference concentrated now mainly on the control of distribution, As a basis for this control, the 1925 Convention, introducing far-reaching innovations, provided: (i) the compulsory system of import certificates and export authorizations and (ii) a comprehensive system of statistical control of all narcotic drugs transactions. The operation of that system was entrusted to the newly created Permanent Central Board (PCB), the first international organ of narcotics control endowed with rather wide powers.

The 1925 Convention came into force only in September 1928. In the meantime the illicit traffic and narcotic drug abuses were spreading on an unprecedented scale threatening large sections of the population in several parts of the world. The Tenth Assembly of the League of Nations (1929), impressed by the alarmed public opinion, unanimously decided to direct the League's Advisory Committee on Opium to elaborate a draft Convention for a direct quantitative limitation of the manufacture of narcotic drugs. This draft served as a basic document for the Conference on the Limitation of the Manufacture of Narcotic Drugs (Geneva, 27 May to 13 July 1931) which concluded the in many respects remarkable Convention of 1931 for Limiting the Manufacture and Regulating the Distribution of Narcotic Drugs. Supplementing and completing the 1912 and 1925 Conventions, the 1931 Convention gave full effect to the general principle of the limitation of manufacture and consolidated the control over both the manufacture and the distribution of narcotic drugs. The 1931 Convention introduced the compulsory estimates system and through it the quantitative limitation of the manufacture of and trade in narcotic drugs. The Drug Supervisory Body (DSB) has been established to administer the estimates system.

Under the compulsory estimates system, which is now being continued, with an enlarged scope of application, under the 1961 Single Convention, all governments of all States in the world-Parties and Non-Parties alike - were requested to furnish estimates of the quantities of drug required for legitimate needs in their respective countries and territories during the coming year. In the case of any government failing to furnish these estimates, they were established for the country or territory in question by the Drug Supervisory Body (DSB) - now by the International Narcotics Control Board (INCB) under the 1961 Single Convention - and they were (and continue now to be) in every respect as binding and having the same legal and administrative effects as if they were furnished by the government itself. A consolidated annual Statement showing the estimates for each country and territory in detail is sent to each Government in the world. This Statement is in fact the world plan of drug manufacture and distribution. It is binding on all governments in their mutual relations.

The quantitative limitation of manufacture of and trade in drugs was thus ensured considering that no greater quantity of any of the drugs shown in the Statement could be legally manufactured in any one year in any one country than the total of the quantities necessary for executing export orders in accordance with the provision of the 1931 Convention. A non-manufacturing country was not authorized to import in any year a quantity of drugs exceeding the one shown in the Statement plus a quantity needed for re-export. If it did, an automatic embargo notified by PCB to all countries stopped the exports to that country of drug imported in excess of that country's estimates.

The results of the full implementation of the 1925 and 1931 Conventions were highly satisfactory. Unlimited manufacture in licensed factories and free trade in narcotic drugs came to an end. One of the main objectives of the efforts that began in 1909, namely the drawing up of a clear-cut dividing line between the legitimate trade the illicit traffic, had been achieved. The licensed factories ceased to be the source of narcotics for non-medical purposes.

A few figures, showing the movements of the main drug:

morphine, will illustrate this change. Taking into account the fact that not all manufacturing countries have furnished complete statistics of morphine manufacture during the 6-year period 1925-1930, the minimum annual average of authorized morphine manufacture during that period, calculated on the basis of these incomplete data, amounted to 44.3 tons. The maximum legitimate world needs of morphine for the same period of six years were estimated by the Secretariat of the L.O.N. at 29 tons per year. Thus on the average a minimum of 15.3 tons escaped each year from licensed factories into the illicit traffic; a total for six years: 92 tons. During the period 1931-1935 (i.e. after the entry into force of the 1925 and 1931 Conventions) the average annual manufacture of morphine was stablized at the level of the world legitimate needs, namely at 29 tons. The figures showing the authorized exports of morphine are also of interest: these exports have reached in 1926 12.3 tons and decreasing progressively they stood at 1.17 tons in 1937 (or at 9.5% of the exports in 1926).

The flow of drugs from licensed factories having dried up, it was replaced by a stream flowing now underground from clandestine manufacture.

Clandestine laboratories and factories sprang up in different parts of the world establishing themselves in areas of weak governmental authority or currupt and simply non-existing effective administration, but mostly in proximity of uncontrolled or illicit opium production. This "free" production of narcotic raw materials and clandestine manufacture of drugs had to be brought under control. The work of the League of Nations on a draft Convention that would subject these raw materials to quantitative limitation (comparable to that existing for manufactured drugs) was interrupted by the Second World War and resumed only in 1946 when the UN undertook to continue the League's efforts in this field.

After more than six years of negotiations and preparatory work of a Draft Opium Protocol submitted by the UN to a Conference of Plenipotentiaries was signed in New York on 23 June 1953; but it took another 10 years to put it into force (on March 8, 1963). This, in the history of international treaties on narcotics, excessively long delay of putting the 1953 Opium Protocol into operation, illustrated once more the differences between planning and regulating, 1) industrial manufacture and 2) agricultural production.

Admitting some exceptions for a limited period of transition the 1953 Protocol prohibited consumption of opium for non-medical purposes; introduced the system of annual estimates of opium production and opium requirements "with a view to limiting to medical and scientific needs the quantity of opium produced in the world;" limited the amounts of stocks of opium which each contracting Party is authorized to maintain; it required the establishment of national opium monopolies in the opium producing countries; enumerated by name seven countries which alone are authorized to produce opium for export; and increased substantially the responsibilities of the PCB making it the main international organ for the implementation of the 1953 Protocol in the international sphere (e.g. it empowered i.e. the PCB to take against a country committing a serious breach of the protocol the severe sanction of the mandatory embargo on the export or import, or both, of opium).

The 1953 Opium Protocol was the 9th international instrument dealing with narcotic substances. It was now a practical necessity to simplify this international legislation by consolidating the existing treaties in a single instrument. This work has been concluded by the adoption of the Single Convention on Narcotic Drugs, signed at New York on 30 March 1961. Coming into force in December 1964, this Convention brought under control <u>all</u> narcotic substances. It outlawed their production, manufacture, trade and use for non-medical purposes; it limited possession of all narcotic substances to medical and scientific purposes and to persons authorized to possess such substances; it extended the estimate system of the 1931 Convention to all narcotic substances and stipulated that the system of import certificates and export authorizations has to be applied also to poppy straw (used as raw material in the manufacture of morphine); by incorporating the basic provisions of the 1953 Protocol, the 1961 Convention provided for the international control of all opium transactions by what are in practice national opium monopolies ("government agencies"), authorizing opium production only by <u>licensed</u> farmers in areas and on plots designated by these monopolies.

Under the 1961 Convention, the main international organ for the implementation of that Convention in the international sphere is the International Narcotics Control Board (INCB). Replacing its predecessors, the PCB and the DSB, created by the 1925 and 1931 Conventions respectively, the INCB inherited their responsibilities which were increased by new ones under the 1961

Convention. The new Board, in order to exercise properly the functions of an independent judiciary organ of the international system of narcotics control and to meet its treaty obligations has:

(i) to administer the estimates system as well as the system of statistical control of all transactions concerning narcotic substances under the 1961 Convention;

(ii) to watch continuously the course of the international trade in narcotics with the view to preventing any country from becoming a centre of illicit traffic;

(iii) to investigate matters relating to narcotic substances which may be brought to its attention under the provisions of the Convention by a Government or by United Nations organs;

(iv) to seek explanations in respect of apparent breaches of the treaties;

(v) to bring the matter of treaty breaches before the Economic and Social Council of the United Nations;

(vi) to arrange for the publication of its reports; this confers upon the Board the right of addressing itself to the public opinion of all countries; the Convention stipulates expressly that "the Parties shall permit. . . . unrestricted distribution" of the Board's reports (Article 15);

(vii) to recommend in situations in which the aims of the 1961 Convention are being seriously endangered by any country, Party or Non- Party to the Convention, that Parties "stop the import of drugs, the export of drugs, or both, from or to the country or territory concerned, either for a designated period or until the Board shall be satisfied as to the situation in that country or territory" and to call upon a government "to adopt such remedial measures as shall seem. . . to be necessary for the execution of the provisions" of the 1961 Convention.

If I were asked to sum up very succintly this rather fragmentary presentation of some aspects of the international system of narcotics control, I would mention as its main results that:

(i) the amounts of narcotic drugs legally manufactured approximate now very closely the world's legitimate needs;

(ii) the escape of narcotic drugs from licensed factories has ceased; and

(iii) the existing illicit traffic in these drugs is drawing now its supplies from clandestine factories;

and I would add that :

(iv) the elimination of the clandestine manufacture depends largely on the suppression of the illicit and uncontrolled production supplying narcotic raw materials to clandestine factories;

(v) efforts should be now concentrated on that task which, considering its complexities and serious difficulties, require different methods and more time than the bringing of manufactured drugs under effective control.

In retrospect, reviewing the evolution of efforts during the last 60 years and the series of international treaties concluded during that period - a historic period of deep, global, often violent changes of human institutions and conditions of human existence - the journey that started in 1909-1912 appears as a continuous voyage of discovery, requiring at each bend and turn new efforts to avoid pitfalls and danger points, new measures and techniques to overcome new obstacles and constant concentration on both the immediate tasks at hand and the final goal.

The tasks at hand can be dealt with adequately and the final goal reached only if we are constantly aware and, if necessary, insist repeating that "international action is a supplement to, not a substitute for, responsible action" on the part of national governments. *

---

* Quoted from "International Aspects of Drug Abuse Control" by Harvey R. Wellman in the Department of State Bulletin of May 17, 1971, pp. 639-647.

PREVENTIVE APPROACHES TO DRUG DEPENDENCE - EDUCATIONAL APPROACHES

Akhtar Husain

Programme Specialist, Department of School and Higher Education, UNESCO, Paris

The drug problem is basically a problem of control of demand and supply of drugs; it is a question of creating fool-proof mechanisms for production and distribution of drugs in permissible quantities, for legitimate uses and eradication of their misuse or abuse. But experience has proved that there is nothing foolproof in an imperfect and complex society. And therefore, while efforts for improving these mechanisms through more rigorous and effective legal and administrative measures continue, we are being forcefully reminded of the good old saying: Prevention is better than cure.

The importance of preventive measures to check abuse at its very root is being recognized and emphasized since the adolescents and the young, in many developed and developing countries, after the Second World War, have started indulging in drugs. Until then,the problem was confined more or less to the backwaters of society and was considered at worst as a nuisance to be handled by the police and the doctors.

Now that within two decades young people in and out of school, in increasing numbers, from poor as well as rich homes, for different reasons have decided to experiment with old and new drugs, and in fact claim to have created a drug cult or a drug culture, their elders are alarmed and baffled and are trying desperately to stem the tide. Naturally they expect the educational system to play an active role in the solution of this problem.

It is a reasonable expectation and the educationists, caught by surprise as all others, are applying their minds to the dimensions of the problem and the best approaches to meet it. Their task is not confined to the school system only, because the primary objective of a drug education programme is to present basic instruction about drugs and their effects and to create awareness of the multitude of factors causing people to use, misuse and abuse drugs and narcotics. In this sense education is to be given a broad meaning and should be a vehicle of intelligent information rather than an instrument of narrow instruction.

Thus, the drug education programme has to address itself to various categories of drug takers; first and foremost to experimenters and occasional users who are still young and much more numerous than drug-dependent persons.

What should be the substance of this programme, and the means and agents of its execution? No one can give clear and categorical answers to these questions, because of the very nature of a hydra-headed problem. Therefore, drug education programmes will have to be evolved and adjusted to requirements of different situations and communities. Certain countries have already started such programmes within the last few years and one has to allow some time to verify their achievements. In this connection, one has to give due consideration to the difficult task of the educationist. He does not know if the mention of drugs to the uninitiated school children will excite their curiosity and attract rather than dissuade them from experimenting with these objects. He is also puzzled by the demarcation line between socially approved and socially disapproved drugs and does not know how to justify such arbitrary demarcation. He does not know how to explain that what is admissible for adults is forbidden for adolescents. Then, there are conflicting views among experts on effects of some drugs, cannabis for instance, which is used by millions of people. What is the teacher to advise in this case? No education programme can succeed without the cooperation of the student community, and the basis for such cooperation is to be established by a free and frank dialogue between the educator and the student, whether in school or out of school.

The problem is so new, sudden and explosive, that the ground for any effective educational approach to drug prevention is to be prepared with great care. While the premises for action will be determined by every country according to local requirements, certain activities of a common character can be identified and implemented under international auspices which are best suited for ensuring action on a multi-national scale.

To begin from the very beginning, no one knows exactly what is being done under drug education programmes in different countries and to what effect. And yet, a good deal of experimental work is being done although hardly any evaluation has been done. It is necessary to make an objective evaluation so that worthwhile experience in one area is brought home to other areas. Similarly, a careful survey of the extent of drug problems among the youth of different countries is urgently called for. These two assessments will provide essential background material to study the extent to which drugs have crept into the life of the youth and what is currently being done and planned to meet the problem in the educational field. This leaves out the important question, why the young are attracted to drugs. It is primarily a matter for social science research. Some research has been done, but it is dispersed and there is no way of taking stock of it. It is important to survey the scene and chart out further avenues of research. These surveys will be useful for all those who are engaged in the solution of the drug problem, particularly those who carry the responsibility for planning and executing drug education programmes.

Granting that the framework of a programme is established, trained personnel for its execution will not be available because very few educators have ever dealt with the problem. This does imply the need for their training through seminars, study trips, etc. At the same time, educational materials, both for teachers and students, will have to be prepared.

So far, we have discussed the necessary tools for an effective drug education programme, leaving out the question of the programme content. It is obvious that it will differ from country to country, community to community. But a few facts will be patent to every situation. Existing school curricula offer many opportunities for the introduction of drug education and the educational approach will have to be interdisciplinary. For example, problems of drug abuse can be introduced into health and nutrition education, social studies, natural sciences, etc. It can take its points of departure from appropriate contexts in such studies and from the interest of the pupils themselves. A good deal of tact and imagination is expected in preparing materials and in training personnel because these are the twin pillars of drug education programmes like any other educational programme.

All countries which have a drug education programme work with the school system, either through the Education Ministry or with local schools and adult education centres. These efforts range from the dissemination of materials to the training of educational personnel. In the United States of America and in Sweden where considerable experience has been gained, the materials consist of: guides on methodology, audio-visual slides and tapes, booklets for self-instruction, bibliographies, etc. Teachers trained are from biology, social science and home economics.

This paper does not venture into the philosophy of the drug cult or the civic and health reasons in its refutation. There is a plethora of literature on the subject in defence as well as rejection of drug use, and those interested in the subject are aware of the arguments for and against drug indulgence. Our discussion is limited to the exploration of educational action to help in the prevention of drug abuse at the stage where it may start or may have started so that the mental health of the user is not impaired. It will be useless to pretend that a drug education programme, however efficient, will by itself do the trick. Much will depend on the co-operation not only of pupils but also of their parents and society at large. Every problem is inter-connected with other problems, and the educational approach to this problem will have to be supported and co-ordinated with other efforts to produce the expected results. As the roots of the problem are tangled in sociological and psychological factors, they will have to be attacked at local, national and international levels on inter-disciplinary and inter-agency forums. Education, in a broad sense, can play a constructive role in this undertaking.

DR. EVANG: Thank you Mr. Husain. I think you were all too modest when you introduced your paper. It was very useful indeed. I am not going to try to summarize the contributions which have been made. We were about half an hour late in starting this session. One comment on the paper by Dr. de Moerloose: I have been asked by one or two participants to invite comment this afternoon on two aspects of legislation which have a bearing on prevention. Firstly, is it of any value, or is it not, to undertake registration of drug dependent persons? Secondly, what is the value or non-value in preventive effect of compulsory treatment? I would like you to think this over during the lunch hour. As far as international legislation is concerned, of course, we can sense a trend towards harmonization which is most helpful and which I think is encouraged by the activities of the World Health Organization and other international bodies in this field. Some of us feel this process is slow - too slow in fact. There are various reasons for this, limitations based on the different situations in the various countries. Among other things, when we speak of legislation in the field of prevention or touching upon prevention, countries differ very much in the way in which they have organized their social and health services. This has to be taken realistically into consideration when you are drafting your legislation. For one, I feel sometimes that perhaps the lawyers are not quite up to this situation. As far as the international aspect is concerned, I would remind you that that country which pressed very hard originally for the first opium convention - the United States of America - is now again pressing very hard to establish new and stricter rules in the international field, and certainly we are hoping that this will produce a result. Let me also say we hope that this long period of years has proved the necessity for what you might call international discipline and sacrifice in this field. No individual country should allow itself to take steps based on its own problems only, without taking full recognition of the direct and indirect effects this might have on other countries. I would welcome in this group - with wide experience - also remarks on this point.

Now to Mr. Husain's paper. I said that you had been much too modest, and by that I meant that in fact you made five significant points. Firstly, you underlined that educational programmes should be evolved and adjusted to the different situations and audiences. This is a very important principle. Secondly, you underlined the necessity for those who are taking part in education that before they do so, they should really make up their own minds on the controversial questions which may be expected. Otherwise when the time comes for the controversial questions or the provocative questions to be discussed, they may find themselves in a situation in which they are unable to answer. Everyone should be able to give his opinion freely and you cannot do that on an impulse - you can only do it if you worked with yourself for quite a long time. The spontaneous individual testimony does practically never represent the "true" reaction. Thirdly, you have underlined the necessity for building into every educational programme an evaluation procedure. Admittedly this is difficult both nationally and internationally, but we should try. Fourthly, if I understand you correctly, you have underlined the necessity of undertaking this education as far as the school is concerned, as an integrated part of the school activities, not by people coming from the outside, causing sensation. I also highly appreciate the point you made, that information on dependence-producing drugs should be given not only in connection with health or nutrition, but, for example, also as part of social studies. This is an excellent approach. If you use the opportunity to discuss the social, economic, psychological and other aspects with children and youngsters, you immediately have an educational approach which is useful. Finally, you have underlined that, of course, when we speak of prevention, information and education is only one part of the total picture, I think this afternoon we may also discuss this - do we overestimate the effect of education and information, or don't we? I recall the remark of Dr. Freedman, very wise I think it was - who said that human beings cannot be expected to react rationally to substances which they introduce into their own body. I agree fully with that - and that is one of the reasons why education and information have its limits. If you don't know the limit of your weapon, you should be reluctant using it, because your range may be completely wrong. I think we will adjourn for lunch and I will briefly introduce my paper immediately after lunch.

COMMUNITY APPROACHES TO DRUG DEPENDENCE

Dr. Karl Evang

Director-General of Health Services, Norway

**

Since the time approximately a hundred years ago when scientific medicine combined its impact with that of the improved standard of living, we had - at least in the technically most advanced countries of the world - become accustomed to regard it as self-evident and "normal" that further progress towards better health for everybody was safely under way.

There were admittedly exceptions, like heart and circulative disease, accidents in traffic and in the home, venereal disease etc., but this could more or less be explained in words which everyone understood.

No wonder therefore that the sudden appearance and rapid spread of abuse of dependence-producing drugs caused confusion in the community, anxiety, indeed sometimes panic. Here was a new "disease", incapacitating frequently for life a growing number of young individuals. It even began to make a place for itself in the list of causes of death. In New York city there was for example estimated in 1969 that fatalities due to abuse of drugs was a leading cause of death in the age group 15 to 35.

The new disease spread from country to country through channels which to a large extent escaped identification and control and where therefore effective quarantine measures could not be applied.

The richer and more highly technically developed the country, the more fertile the soil seemed to be for the new pathogenic agents. The heterogeneity of the victims were also striking; they came from every thinkable social class, group or profession, with widely varying social, economic, politically religious, educational and emotional background. Younger and younger age groups were involved. Obviously the causation was highly complex. The general public and the politicians could not avoid being confused by the chorus of representatives from many disciplines: Psychiatry, psychology, pharmacology, physiology, toxicology, sociology, criminology, pedagogics and law. Laymen were invited to accept widely divergent views: Some held intoxicating drugs in general as a door to new reality and culture, others felt that the growing drug abuse would topple mankind over the brink into the abyss.

Out of this chaos it gradually became clear at least that one had to deal with <u>a community mental health and social relations problem</u>. The widened definition of health introduced in 1946 by The World Health Organization, including a somatic as well as a mental and social component, was applicable. But although this definition of health had since that time been ratified by practically all countries of the world, had the health - and social services developed accordingly? And what could parents, teachers, youth leaders and others do when the problem actually turned up in the home, class, group?

First and most important: What should young people themselves think? How should they behave in a world where drugs of all descriptions were floating around? Were all these drugs dangerous or could they find a drug which they could try with only negligible or no risk? Perhaps even with advantage?

Community reactions in the preventive as well as in curative (including rehabilitative) fields varied in the beginning widely, were erratic, spotty, sometimes directly confused and panicky. Faced with a new multi-facetted problem which necessarily involved a variety of specially trained professions and which asked for participation of several organized branches of society (health and social services, schools and other educational institutions, parents, police, drug control and customs authorities, etc. besides voluntary organizations and lay groups) it was only natural that it took some time to develop rational approaches. Co-ordination, open discussion, sharing of responsibility and joint action was necessary. Health and social services cannot carry the burden alone, neither can schools, parents, police, custom officials or others. The greatest difficulty seems to have been to make parents understand the vital role they must play and further to make teachers accept their role in early detection and remotivation of young people in the danger-zone preferably already before the maladaptation has given rise to drug problems.

As far as the police was concerned it turned out to be of fundamental importance to avoid a polarization between police on the one side, and social and health services on the other.

Community approaches in the field of prevention to me mean the total programme of - hopefully co-ordinated and hopefully useful - steps to prevent, curtail, control and guide the complex situation in which we find ourselves.

Before starting to organize the health and social services one is generally recommended to develop a philosophy, to define the goal. In this case I would warn against following this rule too literally. A variety of problems obviously exist and cannot be explained away. As soon as one feels that one can at least tackle one of them, one should in my opinion not hesitate to start, even if the philosophy is not fully developed and some of the goals perhaps still somewhat foggy. It takes time to establish service facilities and some problems will have to be solved "in the march". In the preventive as well as in the curative and rehabilitative field it is obvious that the traditional service structures which have so far been developed in health and related social services are not adapted and not adequate to meet the need in this field. A period of trials and errors should be foreseen.

Agenda items 4. 1, 4.2 and 4.4 cover the legal, educational and medical approaches within the preventive field. The delineation from other community approaches is not clear cut, and the following brief comments must be seen in this light. Community approaches in the preventive field varies a great deal but gradually a pattern covering at least the following points seems to be under development in countries with the longest time of exposure to the new type of problems:

1. Systematic attempts to offer children and young people

attractive alternatives for meeting their demand for activity contact, responsibility, acceptance. Such attempts should in anyone area be based on a study of the motivation of children and young people to turn to experimentation with dependence producing drugs in their area.

To find such positive alternatives imply not only to study the motivation, but also to dig deeper and to face squarely the social, economic, and ideological factors at play. Admittedly, this will immediately involve you in tackling problems outside the scope of health - and social services, and the activation of and co-ordination with other forces (referred to below) is necessary. Young people and children should be invited to participate in the planning and running of such alternative outlets for activity, contact, etc., including children and young people who have themselves been undertaking experimentation with dependence-producing drugs, or are on their way to develop a state of dependence, on the condition, of course, that they are motivated for substituting this behaviour pattern with the positive alternative.

In this connection parents, teachers, youth organizations, employers etc. should be alerted and educated for early detection or signs in children and young people indicating specific or general maladjustment which may as a next step lead to experimentation with drugs.

It should be emphasized over and over again that experimentation with dependence producing drugs may be a relatively late sign of maladjustment, although though this is not always the case.

2. More specific methods for early detection ("secondary Prevention") should be developed in and around "danger zones". Continuous attempts should be made to get in touch and establish human relationships and positive dialogues with young people who are to be regarded as highly exposed contacts. Such night or day "patrols" have given considerable dividends, especially if they have at their disposal facilities for offering the young individual in question other alternatives, accommodation for a few days in emergencies and economic means for offering positive alternatives. These early detection activities ask for a team-work of social psychiatrists, social workers, psychologists and pedagogues with special insight in this field. The activity would most

profitably be under the technical responsibility and administration either of the mental health services or the social welfare services. The experience with "enthusiastic" amateurs working in the field are in most cases not convincing.

Multiprofessional teams working directly in the field have to strike a difficult balance: On the one side they must be trusted by the clientele they are trying to approach and assist, on the other hand, they also need the trust and co-operation of those organs within "the establishment" which have a duty to perform in the same field. In a way they represent a link between these groups of children and young people on the one side and the more traditional establishment on the other. They have to meet their clientele, humanly speaking, on an equal level and to a certain extent on their conditions, without being at the same time permissive, in the sense that they blunt their own efforts and destroy their own goal. The more they are able to activate parents, teachers and others directly in their work, the better the chance of success.

3. It is essential to avoid the establishment and continued life of milieus - especially in larger cities - where drug dependent individuals can live for longer periods of time under more or less constant drug influence. Such groups or milieus have a strong recruiting effect on young people who may originally not be interested in drugs, but are motivated by the natural need for young people to establish closer contacts with groups willing to accept them unconditionally.

No large city, at least no capital in the Western world may, I think, now be said without problems of this type. Various types of community reactions - partly planned, partly spontaneous have now been observed for a sufficiently long time to state definitely that a too permissive or laissez-faire-attitude does not solve the problems, but has a tendency to end in minor or major catastrophes. On the other hand, it is obvious that a one-sided condemnatory and discriminating approach has little chance of producing control.

There are, to my mind, two <u>basic conditions</u> for the development of mass abuse of dependence producing drugs. These two conditions are of such a fundamental character that no programme of prevention should be planned or executed without taking them both fully into consideration.

One is that such drugs are available, the other that the use is socially accepted in smaller or larger groups of population.

4. It follows therefrom that restrictions in availability and avoidance of social acceptability should form a most important and fully integrated part of any preventive programme, and I am not going to elaborate this point here. It should be remembered that there is not always a direct relationship between the availability and the actually existing number of misusers. Some countries in the Western world have been reporting that certain types of dependence producing drugs are, in fact, available, in the sense that they exist in the country concerned, but that nevertheless the number of pople who experiment with or abuse these drugs is relatively small. This seems especially to be the case where the country concerned serves as a transit country for such drugs to a neighbouring market where the demand has been sufficiently developed and the purchasing power exist. Whether such a situation is only of a transitory character or may last for longer periods of time, remains to be seen.

Various aspects of restrictive measures to make dependence producing drugs as unavailable as possible have been discussed under other items on the agenda.

Stricter control with doctors' prescriptions of all types of dependence producing therapeutic drugs are important in this context. There is every reason to believe that the over-consumption of psychotherapeutic drugs in middle-aged and older age-groups and the general trend towards manipulating one's mind with drugs, is a contributing factor towards experimentation and abuse in children and young people. Also it is well known that uncritical prescription by doctors may lead to not inconsiderable amounts of such drugs finding their ways into grey or black markets. Removal of certain types of combined (composed) therapeutic preparations may be helpful in this context. Where distribution systems permit, encouraging results have been achieved through automatic data processing of data in regard to prescriptions of dependence producing drugs by the individual doctor and/or consumption by the individual patient. The highly mixed experience with methadone-treatment of heroin addicts should be noted. Several countries now also report an increasing number of primary methadone dependent individuals. Most unfortunately, methadone has in some areas got the reputation of being a drug which "cures" drug dependence in a painless and effective way. As so clearly stated, i.e. by the WHO Expert Committee on Drug Dependence in its 18th Report (page 23) "methadone maintenance is a continuance of drug dependence of morphine type".

5. Any form of social acceptance of dependence producing drugs should be strongly counteracted, generally as well as in "sub-cultures". To organize this on a community basis is, however, a very complex task, involving, i.e. mass-media over which one has little control. A great deal of confusion has also been created through the lack of coordinated public approach by people with varying professional background. The sharp distinction between so-called "soft drugs" and "hard drugs" should also be avoided. While from the purely pharmacological and biochemical viewpoint criteria may perhaps be developed to undertake a more or less clear distinction between the "soft" and the "hard" drug, I have not been able to find evidence to support such a distinction when we discuss drug dependence in the wider and more realistic psycho-social context. I think, for example, it is correct to state that in most Western countries there is a larger number of individuals who have been introduced per year to morphine (for medical reasons) than the number who have started experimentation with hashish. Although in traditional pharmacological terminology morphine would be characterized as a "harder" drug than hashish, nevertheless, in some Western countries at the present time, hashish seems to have produced a larger number of people in a serious state of drug dependence than morphine.

The "deglamourizing" of the drug of choice, which will frequently occur after some time's use should be actively explored and encouraged. Mobilization of young people for this and related tasks is essential. The credibility of the older generations and "the establishment" is limited. Politically and otherwise very active groups of young people have in some countries been the first to "clean" themselves of drug use, having learned by experience that abuse of dependence producing drugs is incompatible with the discipline and activity asked for. It follows logically from what has been stated above that no form of "flirtation" with dependence producing drugs, even the "softest", should be tolerated amongst the personnel in the preventive service facilities, (and that also should be a rule during the curative and rehabilitative phases).

6. Co-ordination at local, regional and central levels should be established between organs and personnel working in the health, educational social, police and customs control fields. Amongst the duties of the co-ordinating bodies would be the promotion of surveys for following as closely as possible development, further reciprocal communication, stimulation and guidance in development of preventive measures. Too rigid rules in the field of professional secrecy may have to be modified to make fruitful co-ordination possible. While in some countries lack of understanding and co-operation - indeed polarization - is most strongly pronounced between social and health services on the one side, police on the other, in other countries schools and educational institutions as a whole have been the most reluctant to enter into fruitful co-operation.

7. Community approaches to preventive activities may fruitfully be viewed within the framework of an epidemiological model. It seems to clarify the position for laymen, politicians, etc. if one defines as clearly as possible:

(a) the pathogenic agents,
(b) the environment, and
(c) the host.

The term "epidemiological model" in this context may need some clarification.

Unnecessary misunderstandings and unfortunate emotions have sometimes been created when the term is interpreted to mean a too close parallel to the traditional communicable diseases. Obviously, there are parallels, and epidemiological terms are convenient to describe certain aspects of the phenomenon under discussion here. One may, for example, speak of an endemic area for drug dependence, of epidemics or episodes, even the term pandemic may be justified. Emotions, making a detached discussion difficult, seem, however, frequently to arise when one goes a step further and speaks of a drug dependent individual as a contagious case, surrounded by contacts, who are strongly exposed. Less objectionable perhaps is to speak of individuals who do not themselves use the drug, but distribute it for profit or other motives, as carriers.

Obviously, the epidemiological model used in this more traditional and restrictive sense has its limitations when applies to the field of drug dependence. When we speak of an epidemic communicable disease, we are, for example, accustomed to look for one specific, identifiable, contagious agent as the dominant cause. Faced with mass "outbreaks" of drug dependence in children and young people, one is on the other hand faced not only with a large variety of dependence producing drugs, but, as stated so frequently, also with highly complex, multiple, psychological, social, ideological factors, differing not only from country to country, but also within the individual country from group to group, even from individual to individual. The "causes" of an epidemic of drug dependence in a slum district in a large city differ, of course, greatly from the "causes" for using perhaps the same dependence producing drug in well-to-do university students. Also we know that a state of drug dependence does not, like most communicable diseases, produce immunity. Quite to the contrary, a person who has developed a state of dependency, will - regardless of the type of drug - in most cases be more susceptible to other dependence producing drugs. Simple isolation to cure a person suffering from a condition where social maladaptation is a major component, is also in fact a "contradicto in adjecto".

What may therefore usefully be stressed is that the epidemiological-statistical method is highly applicable to a multifactorial causation with which we have to deal when speaking of nonmedical use of dependence producing drugs. Some leading epidemiologists in the world today, in fact, occupy themselves more with noncontagious diseases and conditions, like heart infarction, traffic accidents, even child births, than with communicable diseases.

Under any circumstances, it is most important to try to define populations at risk, vulnerable groups, highly exposed contacts, etc.

As far as the population at risk is concerned, two different schools of thought seem at present to exist. It is felt by some that the population at risk as far as children and young people are concerned, is in fact rather limited, at least when one speaks of the risk for developing a serious state of dependence. A number of preconditioning factors are named as more or less important: Broken homes, unbalanced personality, family conflicts, early frustration, highly damaging milieus, etc. Others take the view that as far as children and young people are concerned, the whole population is at risk, at least in countries where dependence producing drugs are available and where they are socially accepted to some extent. Personally, I have come to the conclusion that the safest approach is to accept the latter view. Even under favourable conditions, socially, economically, and in the family, it seems to me that also the "normal" child or young person might find himself in such a degree of uncertainty, with such lack of self-identification, in such desperate search of contact, or with such need for demonstrating to himself that he would dare to run a risk, that an experiment with a dependence producing drug would be a tempting alternative.

Regardless of whether one chooses one or the other of these platforms, one point remains which to my mind represents an inescapable fact, speaking from the view of prevention:

An individual who is undertaking experimentation with a dependence producing drug, belongs eo ipso to a population at greater risk than a person who does not undertake experimentation. It is, of course, a great consolation to us all that out of, say, one hundred experimentators only a small number - we do not exactly know how many - will in due course develop a state of dependence. Most of the experimentators do, as we know, not develop such dependence and discontinue their use. By starting experimentation one moves oneself from a group at low risk into a group at higher risk. From the epidemiological and public health point of view one is, therefore, lead to regard it as one of the most important preventive activities to reduce the group of experimentators as much as humanly possible. This means in fact also to counteract "liberal" attitude towards experimentation.

8. The family, the school, etc. and the community at large should regard itself as a "therapeutic community". The trend towards overestimating the somatic, and underestimating the mental and - especially - the social components, must be

overcome. No real progress can be expected unless one also overcome the trend towards discrimination, moral condemnation and general devaluation of the drug dependent individual. To find the balance between a primitive, condemnatory attitude on the one side, permissiveness and a "laissez-faire" attitude on the other, is perhaps the most difficult task in organizing community approaches. Parents with drug-abusing children form in this respect a special, highly emotional group in key position and should therefore be singled out for information and active involvement in preventive measures.

** These comments, on request, are limited to the situation in technically more developed countries and predominantly refer to problems created by the use and abuse of dependence producing drugs in children and young people.

DR. EVANG: We will now hear the paper of Professor Kielholz on the medical approaches to dependence producing drugs.

MEDICAL APPROACHES TO THE PREVENTION OF DRUG DEPENDENCE

Professor P. Kielholz, M.D.

Direktor, Aus der Psychiatrischen Universitätsklinik

Basel

In any form of disease it is possible to deduce the appropriate prophylactic measures from the aetiology of the condition. The genetic factors involved in drug dependence are always of a multifarious nature, originating in the individual's personality structure, early childhood, social milieu, and ease of access to drugs. Drug dependence is therefore mainly based on the triad of personality, drug, and society.

Medical measures for the prevention of drug dependence can be classified as follows:

1. Prevention of a disposition to drug dependence.
2. Information and enlightenment as aspects of education in psychohygienics.
3. Advisory centres for persons at risk from the abuse of drugs.
4. Prospective epidemiological research.
5. Control of drugs used in medical practice (to combat over-prescribing and polypragmatic medication).

1. PREVENTION OF A DISPOSITION TO DRUG DEPENDENCE

A large proportion of the people dependent on drugs originate from a disordered environment, from disturbed or broken homes, or from families in which the father or mother are addicted to alcohol or drugs or in which there is a history of suicide;

alternatively, they consist of individuals who experienced frustration as children because their parents were separated or divorced, because they were brought up under straitened circumstances, because they grew up outside the family circle, because their mothers went out to work during their early childhood, or because there was a lack of security, confidence, communication, and ideals in their home. Later in life, the repercussions of such childhood deficiencies manifest themselves in the form of lack of self-confidence, hypersensitivity, anxiety, feelings of inferiority, aggressiveness, and - in situations of stress - in the form of a tendency to escape from reality. Only in a small percentage of cases does one find evidence that a drug dependent person was over-protected and spoilt by the parents during childhood. Depending on the individual's disposition, drugs tend to be taken at times of increased psychic stress in an attempt to "correct" or forget reality or to escape into a world of illusion. The first biologically critical period is puberty. Owing to increasing sexual precocity, coupled with retarded emotional development, adolescents of today are confronted with far more tensions and conflicts than in the past. Biologically they now grow up earlier and therefore feel socially dependent on their parents for a longer time. Both these factors lead to an intensification of conflicts between the generations, characterised by aggression and opposition on the part of the young against all forms of authority.

Many scientific papers have been published dealing with the personality structure and social milieu of drug dependents and with the factors which trigger off drug dependence. Examination of young people who did not become dependent on drugs although they had taken them for some time, has shown what a strong influence is exerted by family life during childhood. Particularly important in this connection is the question of many of those who have consumed drugs over a certain period are able to stop taking them. Our own research has revealed that the factors which trigger off drug taking, as well as the familial situations, differ in the two groups, i.e. in those who manage to cease taking drugs as compared with those who fail to break the habit.

In most of the young people who succeed in abandoning occasional drug consumption, the precipitating factors responsible for drug taking are curiosity, a tendency to conformity, group pressure, the pleasure derived from experimenting, and adventurousness, whereas in drug dependents the motives derive from conflicts, emotional tensions, anxiety, feelings of inferiority, and aggressiveness.

Those who are able to renounce drugs come from homes in which the family relationships are undisturbed. They have intimate contact with their families, they are self-confident and resolute, and they are in close communication with their parents; many of the drug dependents, by contrast, have grown up in broken homes. The most important form of medical prophylaxis therefore consists in affording maximum protection for the family, especially through care of the mother, family-welfare advisory centres, and "family therapy"; the aim should be to ensure a harmonious family life, the security, confidence, and safety of which protect young people against the urge to escape from reality when faced with stressful situations in the course of their premature puberty.

2. INFORMATION AND ENLIGHTENMENT AS ASPECTS OF EDUCATION IN "PSYCHOHYGIENICS"

Education in psychohygienics should already begin in the family and in the elementary school. The behaviour of children when afflicted with any kind of psychic or somatic disorder is strongly influenced by the example their parents set them. Sport, physical training designed to increase their endurance, and information aimed at inculcating a healthy mode of living, help to prevent young people from taking drugs.

The form of enlightenment which children and adolescents receive should be adapted to their ages and integrated into the overall framework of educational programmes. Use should also be made of statistics and documentary material. Academic speeches delivered to pupils *ex cathedra* often have an effect quite the contrary to that intended, and may indeed prompt some members of the audience to experiment with drugs. The problems of drug abuse should be discussed clearly and objectively in small groups. In such discussion groups, questions concerning the freedom of the individual crop up regularly; here, it is possible to point out that drugs not only fail to solve the problems of the person who takes them, but also actually restrict his freedom, because he becomes a slave to the drug and a prisoner of the drug dealer.

Enlightenment proves useful only if repeated often enough and adapted to the recipient's age and if its effectiveness is subject to continuous scrutiny.

3. ADVISORY CENTRES FOR PERSONS AT RISK FROM THE ABUSE OF DRUGS

Advisory centres and the provision of advice through the telephone should be organised for drug dependents and their parents, who should be able to receive the advice free of charge and anonymously, without fear of criminal proceedings. Whatever form the advice takes, it is important that it should contain objective and medically sound information. Any attempt at intimidation is wrong and has the opposite of the desired effect.

It has been found useful to allow rehabilitated ex-dependents to participate in discussions and enlightenment sessions. Young people tend to place greater faith and trust in those of the same age group and with the same experience of drug taking than in doctors and psychiatrists. Drug dependents must also be reassured that they will not fall foul of the law; otherwise they will cease consulting advisory centres. They should, in fact, not be punished, but kept under sympathetic medical and social supervision.

4. PROSPECTIVE EPIDEMIOLOGICAL RESEARCH

Essential for an objectively conceived and timely form of specific prophylaxis are sustained prospective epidemiological research projects: new trends among drug dependents, the spread and incidence of drug abuse, the types of drug consumed, and the personality structure and social background of drug dependents should be clarified on both a national and an international basis using uniform terminology.

National and international research centres should revise and coordinate the countermeasures taken and should compare their results; the research workers concerned should exchange experiences, and every effort should be made to organise joint research on an international plane.

5. CONTROL OF DRUGS USED IN MEDICAL PRACTICE (TO COMBAT OVER-PRESCRIBING AND POLYPRAGMATIC MEDICATION)

Whereas narcotics are subject - under the Single Convention - to uniform international and national control, many countries impose no uniform control over the other drugs used in medical practice. To quote an instance in point, the Drug Advisory Committee of Great Britain recently reported that in London two general practitioners had prescribed no less than 24,000 doses of methamphetamine for 100 patients within the space of only one month. The question of the indications for the prescription of methamphetamine should be re-examined. This drug ought, in fact, only to be employed to treat narcolepsy and hyperkinetic disturbances in children.

Many other types of drug that are liable to cause dependence are also being prescribed in cases where they are not indicated, and are often being given in large quantities. Therapeutic measures should be aimed primarily at the aetiology of a disease rather than at simply relieving its symptoms. Closer adherence to the principle of causal therapy can help both to prevent members of the medical profession from fostering the misuse of drugs and also to combat the increasing trend towards polypragmatic medication.

The primary responsibilities of practising physicians with regard to the problems of drug abuse can be summarised as follows:

1. Access to dependence-producing drugs must be avoided by concentrating on causal therapy and by cultivating cautious prescribing habits.

2. Attempts should be made to enlighten the various age groups by drawing attention to the therapeutic value of drugs and by discouraging interest in the escape from realty offered by dependence-producing drugs.

3. By striving to protect the family unit, and by recourse to "family therapy" and psychohygienic education, children should be prevented from developing a disposition towards the abuse of drugs.

4. It is the duty of all of us to endeavour to minimise human discomfort, grief, social injustice, or any other forms of social harm, and to provide drug dependents with the best possible treatment.

DR. EVANG: Thank you Professor Kielholz. Before I call on Dr. Smith to present his paper, I would like to call on Dr. Titeca of the Royal Acadamy of Medicine of Belgium. Dr. Titeca will show some slides of electroencephalographic tracings in cases of intoxication by psychotropic drugs.

THE CONTRIBUTION OF ELECTROENCEPHALOGRAPHY TO THE DETECTION OF WITHDRAWAL SYMPTOMS- IN CASES OF PRE-EXISTING DRUG DEPENDENCE

Dr. Jean Titeca

First Vice President, Royal Acadamy of Medicine of Belgium.

DR. JEAN TITECA: I am glad that the subject discussed at this Round Table have not involved solely LSD 25, marihuana, hashish and other drugs of the same kind, of which in my opinion there is a tendency at the moment to speak too often. Personally, I am not convinced that the accounts so prodigally offered the public by the press, radio or television are a means of combating that type of drug addiction: I fear that, on the contrary, for some people it is the safest way of advertising drugs, making them known and taking advantage of the ever-awakened curiosity of young persons in order to incite them to taste of forbidden fruit. My intention therefore is rather to speak to you for a few minutes of abuse of drugs reputed harmless, most of which in Belgium can still be sold without medical prescription. These types of abuse, which lead to addiction more readily than is generally believed by the layman or even by some non-specialized physicians, affect a relatively high percentage of our people, particularly among the adolescents and young adults, whereas the consumption of psychodysleptic drugs happily takes only a very limited toll in Belgium for the moment. Furthermore, if any education of the general public regarding possible risks of habituation and addiction is needed, it seems to me that it should cover first and foremost the unwarranted and medically uncontrolled consumption of psycho-sedatives, tranquillizers for the treatment of anxiety, psychoanaleptic agents and anti-depression drugs, particularly since, while there is no morphine addict who is unaware that the repeated taking of regularly increasing doses of his favourite drug is putting his health in danger, most persons addicted to psychotropic drugs have no suspicion of the risks that they are running, to such an extent that very often they do not mention their drug abuse to the doctor who examines them, not because they want to conceal it but because in all good faith they consider that such a disclosure would have no importance for the determination of the treatment designed to free them from the illness for which they are seeking advice.

This introduction is to show how much I share the wish expressed many times here that we had better ways of diagnosing drug addictions, even, and above all, those which result from the abuse of drugs which the layman, encouraged in his belief by the reading of the encouraging literature accompanying the medicament, considers as harmless. You will not be astonished therefore, when I have added that as long ago as 1934, together with Professor Bremer, I was one of the first to become interested in electroencephalography, to hear that I very naturally extended my investigations to the bioelectric activity of the central nervous system of drug addicts.

A great deal could be said on this subject, particularly regarding the bioelectric syndrome of drug saturation, which manifests itself in many cases in electroencephalographic changes consisting mainly in a slowing-down of the dominant frequency of the base rhythms or in the onset of rapid rhythms, or yet again in an increase in the number of diffuse slow waves. I could also speak of changes in cerebral electrogenesis accompanying the clinical habituation syndrome and also changes sometimes resembling irritation which confirm that the patient has reached a stage of chronic intoxication characterised by the usual constellation of clinical signs, which suggest unhappily that physical and psychic dependence is already established. However, it is mainly in the objective examination of the abstinence syndrome that the use of electroencephalography has proved most effective. In the case of barbiturates, research of this kind had already been carried out successfully first of all from 1953 onwards by a team at the National Institute of Mental Health in Lexington, notably by Professor Isbell who is here, and more recently by Wulff of Denmark who wrote a remarkable paper in 1959 in the Journal of Electroencephalography and Clinical Neurophysiology. Many of you will obviously know of this publication and are therefore aware that persons who have been taking short-range barbiturates in progressively increasing doses over a sufficiently long period show, three to four days after withdrawal, very varied minor disorders but also in some cases much more disturbing neuropsychic troubles consisting on the one hand in the appearance of confusional hallucinatory delirium similar to delirium tremens in alcoholics, and on the other hand tonic -clonic crises of the grand mal type. Now the great majority of these cases of intoxication, even those without such serious complications, manifest between the fourth and tenth day of abstinence really pathognomonic electroencephalographic disorders which arise selectively at the time when the patients, subjected to intermittent flashes of light, re-close their eyes when they have just been asked to open them. At that time there is a discharge of wide slow steep-fronted waves mixed with spikes and sometimes accompanied by myoclonia upon flexion of the upper limbs.

I have applied this technique in examining numerous patients who have abused their psychotropic drugs on a more or less considerable scale - and have not always owned up - with similar results. Through lack of time I shall only be able to comment briefly on some EEG traces recorded in a patient who was addicted to hexapropymate, a non-barbituric sedative sold in Belgium under the trade name of Merinax. The slides I am going to show you come from a research programme carried out in my clinic by Dr. Manuela Utrilla Roblès, who was one of my assistants at the time and has now been working for five years in Geneva in the Department of Professor de Ajuriaguerra and that of Professor Garrone.

The patient is a nurse, 31 years old, immature, with character disorders, who since adolescence has given herself over to various toxic substances to mitigate her difficulties of social adaptation. She has taken successively dolosal, spasmalgine, tuinal and glutethimide. Three years ago she got to know the latest product on the market, hexapropymate or Merinax, of which she was finally taking 40 or so tablets a day. Electroencephalograms taken in the standard way show only moderate dysrhythmia characterised by the presence of short series of rapid rhythms of moderate voltage and quite numerous diffuse slow waves. Furthermore the record shows slow, steep-fronted waves and spikes which are sometimes quite wide. Hyperpnoea only moderately accentuates these bioelectric disturbances, which tend to fall in intensity during the two or three days following withdrawal only to become more intense again for the subsequent ten days or so. However, the important fact is that from the third day onwards flashes of light lead to the appearance of discharges of slow waves and spikes of very high voltage, discharges which are very often wide when the frequency of the inducing flashes is adjusted to about fifteen per second. At this selected frequency stimulation has even had to be stopped after only a half second because the bioelectric manifestations of irritation had straight away taken on such a magnitude that it appeared obvious that to prolong the test would have triggered off a clinical crisis. These phenomena rise to a peak towards the fifth or sixth day and then become attenuated, to disappear between the tenth and twelfth day.

Similar bioelectric disturbances have been recorded since in patients intoxicated by other psychotropic drugs. To keep to psycholeptics alone, I can mention some substitutes for the barbiturates, particularly glutethimide or Doridène, ethinamate or Valamin, ethchloryvnol or Placidyl, acetophenetidine or Saridon, apronalide or Sedormid, diocopiperidine or methyprylone and methaqualone or Revonal. I can also mention anxiolytic tranquillizers such as meprobamate, diazepam or Valium and chlordiazepoxide or Librium. In conclusion I should like to mention further a fact discovered during these electro-clinical researches which have been carried out in my clinic for more than 10 years, viz. that the drug addict who no longer takes the drug and is apparently cured after a few days of abstinence really takes weeks and even months to regain a condition which could properly be called "normal." Indeed I have noted on several occasions that a former drug addict is capable of developing a new electro-clinical abstinence syndrome after again taking the toxic substance, even if he only does so for two or three weeks, provided that not more than two to three months have elapsed between the end of the first intoxication and the relapse. This phenomenon, it seems to me, can only be explained by recognizing that following the first period of abstinence the pre-existing pathological level has fallen below a certain threshold but that the disturbances which had led to that threshold being crossed had not yet altogether disappeared, so that a new intoxication raised the level again in a dangerous fashion. In any case this observed fact concords with that reported in Great Britain by Oswald, i.e. that while during a period of intoxication there is a reduction in the percentage of the time occupied during the night by paradoxal sleep, following withdrawal there is a "rebound," which raises the percentage to a figure very markedly higher than normal; the return to normal requires several months of abstinence and sometimes even a whole year.

DR. EVANG: Thank you Dr. Titeca. We will now call on Dr. Roger Smith from California to present his paper on a community based drug treatment program.

ESTABLISHING A COMMUNITY-BASED DRUG TREATMENT PROGRAMME

Roger C. Smith, D. Crim.

Executive Director, Marin Open House

San Rafael, California

Comprehensive community based drug treatment facilities by the hundreds have been initiated in the United States and Canada within the last five years. They represent a major departure from conventional medical, psychiatric, social and institutional approaches to drug abuse which have been utilized since the mid 1920s. Many communities have begun to realize their responsibility for drug abuse as public interest and level of awareness in this regard has increased. Increasingly, public awareness has been stimulated by the understanding that traditional legal and punitive solutions to drug abuse have not been effective, and may, in fact, be criminogenic. Further, patterns and styles of drug use in the United States have changed rapidly since the mid 1960s, when youthful enclaves of drug use developed; ie. the Haight Ashbury district in San Francisco, or New York's East Village. Drug experimentation and abuse has rapidly spread to previously unaffected urban and suburban communities. Widespread drug abuse by American servicemen in Southeast Asia and elsewhere has led to a search for less punitive and more effective solutions, with readily accessable and relevant pro-programming.

Essentially, the success of any drug abuse treatment, prevention or education programme depends on the following: 1. The trust it inspires among drug users; 2. The accuracy of the information which it disseminates; 3. The competency of the professional, lay, and volunteer staff to deal with the particular set of problems unique to each community and individual; 4. The viability for the drug abuser of the constructive alternative life styles it offers.

The specific programme components of a community based facility may vary greatly from one community to the next. However, most programmes operate on a shared set of assumptions. First, people-helping services cannot be offered in a rigid, aloof, and unconcerned fashion. Professional competence alone is not sufficient to effect individual changes in a population which is characterized by deficiencies in social relationships, and inability to share feelings, and a general disillusionment and cynicism toward the "helping professions". Second, the threat of punishment or actual punishment is not an effective deterrent to drug taking behavior, and may, in fact, be one of the variables which makes drug taking attractive, particularly to youthful experimentors, for whom risk-taking behavior is an important part of the maturation process. Further, involvement in the criminal justice system may stigmatize the drug user to such an extent that drug taking peers are the only available and accepting social group. Third, communities which are properly aroused, educated and organized around the issue of drug abuse possess the resources and expertise needed to locally absorb and deal with this particular form of social behaviour. Finally, most community programmes are committed to systemic change as the only realistic way of beginning to change those social conditions from which drug abuse and related self-destructive behaviour is generated. It is this commitment which most often brings such programmes into conflict with established agencies, and it is out of this conflict that social change can be initiated.

The goals of community based models are as follows: 1. To help neighborhoods or communities deal realistically and rationally with problems related to drug abuse; 2. To serve as a trustworthy source of information; 3. To prevent drug abuse through education and early stage intervention; 4. To minimize the impact of drug experimentation and abuse; 5. To provide effective, accessable and individualized treatment to the individual in the community where he lives; 6. To organize, educate and involve existing private and public social, rehabilitive and medical agencies so that they might become part of a community network of services.

The specific programme components within the community based programme depend entirely on the nature of the population to be served. One such programme in California, privately financed and staffed, with diverse mixture of professionals and non-professionals, with broad community support and involvement, offers the following services: 1. Public education, designed to acquaint as many individuals and organizations in the community as possible with the problems existing in the community, available resources, and direction, in terms of appropriate responses and contributions. More specific educational services are available to professional and para-professional workers in welfare, correctional, probation, parole, educational and law enforcement agencies who deal with individuals having drug related difficulties. 2. Preventive education in the schools is done by scientific pharmacological data, examining social pressures toward and alternatives to drug use. Offered in a non-moralistic and interested way, such services can also serve as a source of referral for seriously drug-involved students. 3. Early stage intervention services are structured to provide immediate medical, psychiatric, or family counseling to individuals and families in a period of crisis as a result of drug experimentation. Included in such services are drug overdose treatment, treatment of acute anxiety reactions, threatened family breakups, educational problems, or involvement with the law. Experience suggests that a treatment team which includes non-professionals, who are often ex-addicts as well as professionals, is most effective. 4. Direct treatment services range from physical detoxification, either on an in-patient basis for those who are long term users to out-patient, medically supervise withdrawal for those with recent history of addiction. Hard core, long term addicts may be candidates for methadone maintenance or referral to a therapeutic community. The problems of dealing with anxiety, frustration, or unbearable stress seems to be best handled in groups where other participants are experiencing similar difficulties. Therapeutic sessions with parents are often indicated when family relations appear to be the underlying difficulty. 5. Families of addicts, particularly those who are unsophisticated about drug abuse, can work through the problems of having a drug dependent offspring with other parents experiencing the same difficulties. This service allows parents to work in conjunction with an on-going therapeutic programme, involving their child, to understand the nature of drug dependency and how to handle individual situations appopriately when they arise in the family. 6. The provision of constructive alternatives to drug abuse may be the more important, yet the least developed goal of most drug programmes, Some programmes restrict their efforts to resocializing the individual so that he might return to society in a conventional or socially acceptable role. Others stress the importance of "alternative highs," which will replace the drug experience. Such alternatives include sensory awareness, yoga, meditation, massage, arts and crafts, or religilous involvement. Still others stress the development of vocational alternatives which are appropriate to ones unconventional life style. In working cooperatively with young drug dependent individuals committed to a "hippy" or radical life style, a programme is often placed in a precarious political or social position by providing assistance to those who wish to remain within that life style.

A community-based facility can operate effectively only if there is broad community understanding, involvement and support. Thus survival depends on the community organizational skills of its executive and the commitment to change which can be generated and sustained.

DISCUSSION

MR PFEFFERMANN (Head, Public Health Division, Council of Europe): What will be done in the practical field especially by WHO and the Council of Europe, and perhaps I could add to what our friend Dr. Cameron has already said. We conducted two studies with the Council of Europe since 1967 - one on the public health field and the other in the field of penal law. We came to the conclusion after two years that we would never reach our end by these methods that are intransically very important but do not bring us forward. Therefore, The Committee of Ministers for the Council of Europe decided to organize together with WHO, a multidisciplinary symposium on drug dependence in March 1972. One important aspect of the symposium is composition. There will be parliamentarians of seventeen European Member countries, governmental experts, experts from International organizations and some national organizations with some experts from non-member countries - Canada, Finland, Holy Sea, Israel, Spain, United States, Yugoslavia. The themes of the symposium are:

1. Patterns of drug dependence in Council of Europe,

member States.

2. Causes and prevention.

3. Treatment and rehabilitation.

4. Legal aspects and control.

The following twelve disciplines will be presented among the experts: pharmacologists; pharmacists; psychiatrists; specialists in other field of medicine; public health administrators; jurists; representatives of police, prisons and customs authorities; youth leaders; social workers; psychologists or sociologists; teachers and mass communication experts. The rules for this symposium have foreseen the following: As ideas regarding drug dependence may vary according to age groups, it is of importance that different age groups should be represented at the symposium.

Now to what should it lead. The main purpose is to arrive at conclusions which would pave the way for the proposal of a series of practical measures within the framework of the intergovernmental work programme of the Council of Europe, to reduce drug dependence in Member states and to increase the effectiveness of prevention and treatment by means of close international cooperation. Thank you.

MRS. J. MOSER (Scientist, Mental Health Unit, World Health Organization): I should like to refer to points made by several speakers: by Dr. Fattorusso, reporting on the work of WHO concerning Drug Dependence; Dr. Edwards in his historical introduction and to what several speakers have mentioned concerning multi-disciplinary aspects.

Apart from the eighteen Expert Committee Reports mentioned by Dr. Fattorusso there have been certain other relevant reports prepared by the WHO Expert Committee on Mental Health. The first such Expert Committee, that met in 1949, recommended that alcoholism should receive early attention. An alcoholism sub-committee was established and held its first meeting in 1950. It was in the preparation for this meeting that the long association with the Organization, of the late Professor Jellinek began. The Committee approached alcoholism as a disease and as a social problem. It pointed out that the public health services could, and

and should, make extensive contributions to meeting the problems involved but that local and social measures related to the distribution and use of alcohol were also of considerable value.

At its second meeting in 1951 the Committee concentrated more on practical aspects of development and treatment programmes.

For many years the section on Addiction Producing Drugs (now the Drug Dependence Unit) considered the problems involved mainly from a pharmacological point of view and dealt with drugs other than alcohol. However, several WHO regional seminars were organized to consider problems of alcoholism with special reference to public health aspects; also a meeting was organized jointly between the two units on the treatment and care of drug addicts.[1]

Problems of dependence on alcohol and dependence on other drugs were considered together by the 14th Expert Committee on Mental Health,[2] whose report stated that these problems "create or contribute to major public health problems and should therefore be of concern to all public health organizations and administrations" and that "WHO should provide further leadership in the development of co-ordinated, multidisciplinary, international research programmes and the stimulation of international co-operation and exchange of information on the problems under consideration."

This report had wide repercussions and was followed by the appointment of a psychiatrist to the Drug Dependence Unit, so that increased attention should be given in their programmes to the human and environmental factors in drug dependence.

In order to follow up some of the recommendations of the 14th Report the two units have co-operated in organizing an Inter-Regional Training Course for National Programmes on Problems of Alcohol and Drug Dependence, which was held 12 September to 2 October this year; Dr. Edwards was our main consultant.

The course was held in three host countries - Poland, Netherlands and England - where consultants, with the help of

---

1 Wld Hlth Org. techn. Rep. Ser., 1957, 131
2 Wld Hlth Org. techn. Rep. Ser., 1967, 363

many organizations and individuals in their own country concerned with problems of alcohol and other drugs, had made extensive reviews of these problems, the existing provisions for meeting the problems and the plans for the future. In these reviews attention was given not only to medical and public health but also to sociological, educational, legal and penal and economic aspects of the problems.

Careful preparations were made in each host country to show the participants how far existing national programmes were able to meet the problems concerned, what were the deficiencies and what further activity was required. A selection was made of persons to discuss these questions and of places to be visited to provide a picture of the situation in a brief space of time.

The 20 participants came from 16 countries in five of the six WHO Regions. Prior to the course they also provided reviews of their countries' problems concerning alcohol and drug dependence and the ways these problems were being met.

During the course attention was given to the nature of problems of alcohol and drug dependence and the response to these problems in different societies and specifically to the kind of information needed as a basis for planning national programmes, the means of selecting, collecting and analysing such information and the ways in which it should be used. The need for continual questioning and evaluation of programmes was underlined.

The response to this course was encouraging. We expect to follow it up and try to evaluate its effect on the participants' activities in their own countries. Next year we shall hope to hold a similar course or seminar for persons from other countries, again focusing on the need for planning and evaluation of services for the prevention and treatment of dependence on alcohol and other drugs.

PROFESSOR REXED: May I first make a comment on one of the points raised by Professor Kielholtz? I think he rightly stressed that it is possible through information and advice from public health authorities to get very good collaboration with doctors. This is the experience in our country too and I think in many other countries.

I would like to comment specifically on the example of central stimulants and the conditions where these can be used. We had to demonstrate that only a very small medical need exists for these substances by taking them out altogether from the list of drugs to be prescribed by doctors, but we kept the possibility of allowing their prescription in special cases, if the doctor gives the name of the patient and the condition. It is interesting to note that during the three years of the existence of this system only 800 prescriptions a year were found necessary for central stimulants (amphetamine, preludine, vitalina type). This is a minute use of these drugs, legally, as compared with their use in some other countries. The condition that doctors think requires this type of medication is narcolepsy, as was mentioned by Professor Kielholz. It is also used in cases of epileptics who are heavily sedated. In no case have doctors asked for a license to use this medication for the treatment of the "hyperactive" child, as it is so frequently done in other countries. It is the opinion of my experts that this condition, if it exists at all, is extremely rare. I think that in some countries the quite normal active child is diagnosed as an over active child, and some doctors even put the label of minimal brain damage on these children. This to me is a highly debatable procedure.

My second observation in this general area would be related to the very useful and interesting paper of Mr. Steinig. It is interesting that since he started his description and discussion from the early Treaties and from the single Convention he limited his discussion to the classical narcotics where much of the illegal trade appears. You might get the feeling from such a description that the whole problem of cutting down the supply internationally is concerned with morphine, cannabis, cocaine, etc. I would like to stress that this is not so. The increasing misuse of synthetically produced drugs of all kinds not only related to opium and morphine but to the central stimulants and the various other groups including sedatives, hypnotics, tranquillizers etc. These are all produced industrially and there is a big international problem in their control. Perhaps not so much their production, because most of these drugs which are illegally used are really legally produced, but their commercialization. Somehow during commercial handling of these drugs they seem to get out of control and they come back from the producing countries to other areas. We see for instance large scale smuggling of central stimulants to Sweden from legal production in southern European and the Mediterranean countries. We see from the same areas these central stimulants starting to spread over, for instance to West Africa where quite large seizures have been made in the last few years.

I think it is very important to keep in mind that if we are to cut down on the supply we will have to deal more and more in the future with a new kind of supply - a supply that comes from synthetic processes in the chemical and pharmaceutical industry and this will be a big problem. If we were pessimistic enough we might even say that if we could close down agricultural production, probably substitutes will appear from the chemists. Indeed they are already here. I do not think we should take such a pessimistic view. We should try to cut down supply everywhere and it is a long fight to do this.

DR. FREEDMAN: I am going back to our original discussion. I want to express my bias on amphetamines. I think I am even in print in 1956 saying I saw little use for them. At least, clinically, I always saw a good deal of danger in them and I took Philip Connell's monograph very seriously, having seen clinical examples of it. I also took the Japanese experience seriously. I say that because I am now going to defend the amphetamines. I had the misfortunate of chairing a panel, summoned in the usual perplexity the government finds itself, when our newspapers reported the abuse of amphetamines in children. So that meant we had to take some very busy experts to review the literature and quiz very carefully those people who had actual responsibility for the care of children and for very disturbed children. There are children whose clinical problems are, to say the least, refractory--if you look at the original literature as this panel did. Scan the scene today and you see that amphetamines have become a new scapegoat--a scapegoat for those people who love intoxicants (such as marihuana), but hate pollutants; some of us in the medical profession are hysterical in imputing toxic effects without stipulating dosage; some of our black minority leaders hate to think of any kind of imputation that their children have any problems--rather, it is society that has the problems. Its a kind of polarization. You have a union of people fighting amphetamines. In fact, I remember a Civil Rights lawyer, working for the Congress investigating amphetamines, who indicated that of course he used "pot"-- "there was nothing wrong with 'pot,' but amphetamines were terrible." So, you begin to see something like the rest of us who believe _our_ alcohol is alright, or like the French chemist whom I know who said that you cannot call wine ethanol, but that other drugs are bad. I think this is bad medicine. I think that our approach to amphetamines has to be very clear and very honest, and I think they do have a use in the treatment of children. You look at the original data of Bradley 30 years ago. The children were not all hyperactive; they were seriously handicapped--both depressed children and other active children with varying diagnoses who responded well when drugs were added to in a therapeutic regimen containing the very best of milieu and psychotherapy. So that to stigmatize the drug or stigmatize the child that might benefit from the drug, to overlook data that children do benefit from the drug, and that their management benefits, is simply to overlook what a few excellent practitioners can do very well. I was very impressed with the fact that many practitioners have never used these drugs and do not encounter such disturbed children. They never encounter the challenge. Well, if we do not encounter the challenge, then what are we going to do with a child who might benefit from these drugs? We have no business stigmatizing the drug or the treatment. We had the ironic experience in the National Academy of Sciences' Drug Efficacy Study, where we found amphetamines efficacious for the treatment of obesity--for two weeks, and only for two weeks--but none of us who had--with regard to evidence--to find it efficacious could say that that was wise treatment. What I would like to say is that there are occasions--they are not frequent, but they are not rare--in which the use of amphetamines is quite useful for children. They are not simply hyperactive children, nor are they necessarily minimally brain damaged children. They are children with neurotic or immaturity problems, or with soft neurological signs, who--after diagnosis and after everything else that you can do for them--seem, with pharmacotherapy, in fact, to mobilize their attention. That is what it--the drug effect-- is; it's not a paradoxical effect. To mobilize their attention with the aid of a stimulant drug can facilitate treatment and bring secondary benefits.

If you have not treated these children, if you have not gone into the clinics, if the reporters in our country have not dealt with these children, if the black minority leaders who have raised a fuss have not done this, then we are irresponsible, I think, in scapegoating the amphetamines. Personally, I would like to see their prescription in general practice cut down enormously; and I accept your point without question. There are going to be 8 to 10, 000 molecules if not more that might affect the mind.

Synthetic drugs are going to be abused and used in a variety of ways, and we must take precautions. But to regulate their supply does not mean that we have to develop an attitude that is fundamentally anti-medical, anti-therapeutical, and anti-scientific. To stop sensationalizing drugs, I think, is terribly important for all of us. I think those of us who are concerned about drug control and misuse should recognize that fundamentally it is the distrust of the physician--that he will not be well enough trained to give a drug. Those of us who are concerned about this, concerned about training good physicians, accordingly have to be very careful about how we present our cautions and misgivings, which I fully share.

PROFESSOR REXED: Yes, I would agree, you cannot control the amphetamines without taking them off the market completely-- I also will testify this that we did this in Sweden just to show the low usefulness of them but I would say that it is an interesting fact that in Sweden, the medical profession was wholly 100% in favour of the thing we did - so we didn't do it in opposition or under criticism of the medical profession.

My second observation would be that it is remarkable that in some countries hundreds of thousands and even millions of doses may be prescribed for a population where according to our medical profession in Sweden the comparative population would probably need just a few hundred or just a few thousand doses. Finally, I do not at all agree with your idea of using amphetamines and amphetamine like substances in the treatment of children of the kind we have discussed. My experts have done just the same kind of scrutiny of the literature that you did Dr. Freedman and there is the verdict that this is not to be defended by scientific results.

DR. EVANG: Let me say that we should not misunderstand this policy which is introduced in my country - to mean that we would prohibit doctors completely from using amphetamines. The only thing we would do is, ask them: What is your indication? And we have every reason to do this. During the German occupation of my country, for example, there were amphetamines used to keep people of the underground movement awake at night. People had to be carried out in the middle of the night through the forests to Sweden or over the North Sea. When the war was over, we found that there was no abuse to speak of, but there was a consumption of amphetamines that we thought was on the high side.

A warning was enough to reduce the consumption - by prescription to one-fifth - 20%.

DR. FREEDMAN: I have no problem in believing you, and I am enjoying the experiment, thinking it heuristic and useful for the rest of the profession. I will remind you, however, if you take any psychoanalytic mental hospital in the United States in 1958 and you reported from it, that the psychoanalysts never required phenothiazine for the treatment of schizophrenia; this would be quite true. And it would be irrelevant to the point that I am trying to make about hyperactive children.

DR. DUBE: I only want to emphasize what has already been said here in the last few minutes. I think the most important point is what Dr. Evang you mentioned in your paper - "..... the strongest possible restrictive measures to make the dependence producing drugs as unavailable......" I neither ask to stop the production of drugs nor try to stigmatize the use. We have already some experience in India in respect of these various drugs- 20 years ago 276,000 kgms of opium was used. Now it has fallen down to .1% because of the restrictive measures. What is really essential for use, is dished out under a licence.

The other point is in respect of compulsion. Several laws are enacted but without consideration of their applicability. For instance, in my country smallpox by law is supposed to be compulsorily notified,treated and it is compulsory for a child to go to school, or we have laws relating to prohibition of child marriages. The enforcement of these laws in fact are never carried out. If laws are made for compulsory treatment of drug habit, I wonder who is going to treat them. The facilities are very limited in several countries. In India there are only about 300 psychiatrists and very few medical men know how to treat. No facilities exist in the villages. Even in countries with adequate facilities all the habitual drug users will be impossible to treat. I wonder whether compulsory treatment would be possible and useful?

DR. EDWARDS: I have listened with great benefit to the communications on Legal Aspects, and these presentations must constitute for all of us valuable resource papers. There are though a number of immediate questions which are stimulated by these broad reviews. One such question relates to that peculiar happening - the law introduced for one purpose but later used for quite other ends.

The danger of such happenings must be watched guardedly. To give an absurd example, the Inebriates Acts of the last century were on one occasion used in England to incarcerate a tea drinker, because his family disapproved of him drinking too much tea. This may seem merely funny - it is not so funny when you get to the Harrisson Narcotic Act. Dr. John Kramer has recently written an important paper - it awaits publication but I think he would not mind me quoting him with due acknowledgement - in which he has analyzed statements made by the American Medical Association just before the Harrisson Narcotic Act came into force. The AMA, at that time, gave constant assurance to doctors that it was not the intention of this legislation to prevent the doctor of integrity from prescribing opiates to his addict patients within the terms dictated by his individual clinical judgement. Within a relatively short time the AMA was, however, either mute or even supporting the restrictive re-interpretation of the law. I will give you a third instance, and would be interested to hear of other examples from around the world. In the 1960's Britain introduced new legislation to deal with the amphetamines which, if you analyse the statements made at the time, had as its intention to aid the police in dealing more effectively with the pusher and the peddler. This law has in the event been used to enforce street searches of the young. I do not believe that this was the original purpose. One aspect of the law which we have to review is, therefore, the law's misuse. Another matter which this morning's interesting papers throw up is then a consideration of the benefits that come from harmonization of the law throughout the world. Once you have uniformity it is much more difficult to get out of step, to break from the others, to experiment, to change. You are caught in the pressures of your peer-group. Linked to this is the notion that no nation should act entirely selfishly in terms simply of its perception of its own problems and its own immediate self-interest. It must recognise that it is a country which lives in the world. Let us then however also turn things the other way - let us also remind the world that countries have individual problems. We can see some tragedies where countries have accepted too readily the international pressures. One final point which these interesting papers make is, I believe, this - the law whenever it is enacted enshrines a stated or a hidden hypothesis. In the minds of the legislatures is the overt or covert hypotheses that this law will accomplish something, otherwise why have a new law. Occasionally you will find that there are several conflicting hypotheses. Again, to quote Kramer's analysis, if you look at the civil committment legislation in California you will find that one of its hypotheses was that it would get addicts off the streets, while another and by no means related hypothesis, was that it would make addicts better. Let us seek to define the hypotheses at the moment when the law is introduced, and let us then ask ourselves how these hypotheses can be validated. I do not believe that this is impossible but it means "before" and "after" research, proper collection of the baseline data. We should see legislation as experiment in public health.

MR. STEINIG: Let me start with a generalization. I believe a good law to be a law which, before it is enacted, pre-exists, so to say, within the minds of the majority of a community. If, e.g. two thirds of a population of a State behave as if the law were already enacted, there exists a fair chance that its implementation will be possible; but if say only one third of the community behaves in the spirit of the law before its existence, the chances of its success are in doubt; it might happen to it what happened to the Prohibition Law in America. So much for the sphere of national legislation. International legislation and the effectiveness of international treaties are of a much greater complexity.

Reference has been made to the uniformity of law. Am I right in assuming that what was meant is the universality of the implementation of a treaty? In order to achieve this universality essential to a fully successful operation of the international control system, the international conferences which have elaborated, since 1912, ten international treaties on narcotics control, had to solve the problem of so drafting these treaties that they could be ratified by the highly developed nations with well functioning administrations as well as by developing nations some of which with rudimentary administrations. Only so was it possible to secure the inclusion of the main provisions of these treaties on as uniform lines as possible in the body of national laws of very different countries. But it is one thing to ratify an international treaty and incorporate its provisions in national legislation, and a very different one to implement it effectively. In some countries, for a multiplicity of reasons, the process does not progress much beyond the ratification. It would take too much of your time to tell you how the international control organs, e.g. the INCB, are trying to over-come these difficulties. I shall say only that their efforts cost time but are not entirely without success.

Professor Rexed very kindly called my attention to my omission of mentioning in my statement the Convention on Psychotropic Substances. I had to limit my statement to a very brief analysis of the ten treaties on narcotic drugs, but during the debate I mentioned the Conference of Vienna which elaborated, in January-February this year, the first Convention on Psychotropic Substances now in the process of ratification.

In addition to the problem connected with the implementation of international treaties and resulting from the dissimularities of conditions in the highly developed and the developing countries, I have to mention, for the sake of completeness, the serious difficulties inherent in the origin of the narcotic substances which had to be brought under control, difficulties that still have to be overcome.

In the beginnings of international legislation concerning narcotics the idea was entertained that one international treaty and possibly one method could solve both the problem of limiting to legitimate needs the production of narcotic raw materials as well as the problem of limiting likewise the manufacture, from these raw materials, of narcotic drugs. As mentioned in my statement, the Geneva Conference of 1924/25 did not succeed in adopting as basis for the new treaty a system of <u>quantitative limitation</u> of these two categories of narcotic substances. Six years later, the 1931 Convention limited the manufacture of narcotic drugs strictly to legitimate needs. The results of its implementation were, as stated before, very satisfactory. But it took another 22 years to conclude the 1953 Protocol from whose implementation it was expected that it will limit, to medical and scientific needs, the quantity of opium produced in the world; 10 additional years were needed to put that Protocol into force (1963). As you see we have been learning our lessons very slowly and in the beginnings we were not farsighted enough to see that the problem of controlling the output of some 62 factories and laboratories manufacturing narcotic drugs in highly developed countries is, in its very substance, entirely different from the problem of persuading and inducing many thousands of peasant families, in several developing countries, producing opium often as their only cash crop, to limit or to give up this production. Time, patience, persistent efforts of the governments concerned, work of international control organs, growing pressure of public opinion have gradually brought about a change in the situation. Today there are only three main producers of opium used for legitimate needs: India, the USSR and Turkey (in the order of the importance of their production). Of the total legitimate opium production in 1970 (some 1156 metric tons) India produced 794 tons, the USSR 226.5 tons and Turkey 51.2 tons. Turkey will cease opium production in 1972. Thus in about one year's time two countries, India and the USSR, will have to meet the world's legitimate opium requirements. The production of opium by these two countries pose no problems for the INCB. It needs to be mentioned, however, that in 1969 Iran resumed production with a view to using home produced opium for the treatment of their opium addicts; in 1970 Iran produced for that purpose about 78 tons.

The remaining problem the solution of which presents considerable and unusual difficulties is the elimination of the considerable, clandestine and uncontrolled opium production, mainly in certain countries of South-East and Middle East Asia. This opium, furnished to clandestine laboratories and factories, is used to manufacture narcotic drugs, mainly heroin, supplied through a chain of illicit channels to the illicit markets of Europe and the USA.

FINAL SESSION

GENERAL DISCUSSION

CHAIRMAN: Dr. Gellhorn

DR GELLHORN: I would like to call to order our final Session. The floor is now open for a general discussion on the whole conference. I recognize first, Dr. Ladewig.

DR. LADEWIG (Swiss Academy of Medical Sciences): Mr. Chairman I would like to refer again to one point which was mentioned in the discussions. Education in the problem of drug dependence requires the collaboration of young people. In an enquiry carried out at Basle amongst a representative sample of the school children population, 1500 persons aged 15-20 years, were asked to state the sources of their information on the drug problem. 4% said they obtained information from private medical practitioners. 3% from their teachers, 5% from lectures on drug abuse and 7% from radio and television programmes. When asked to suggest improved methods of the dissemination of information, 20% suggested discussions in small groups with the participation of ex-users; 43% proposed the cooperation of you in teams of doctors and teachers. This result differs only slightly from those obtained in groups of youth below 15 years of age. A large percentage of the latter preferred lectures and mass media communications. From all of this it is evident that efforts should be made to ensure the cooperation of young people in all drug enlightment programmes.

PROFESSOR JACOB: Mr. Chairman, educational programmes contain among many others, information on the pharmacology of dependence producing drugs. The abuse of these drugs, however, should be considered as a special case, and certainly a particularly serious one, of the abuse of diverse therapeutic drugs and automedication. Education in pharmacology should stress that therapeutic drugs in general, should only be used under certain conditions, often strict, that are at present the subject of extensive scientific investigation and that there are no therapeutic drugs that are non-toxic.

The French text of the presentation of Dr. Smith contains a recommendation - that appears to me, to my regret - too optimistic in the present state of our knowledge. He writes:"To give the individual, in the community where he lives, an efficient accessible and personal treatment." Can a treatment be "efficient" if the community in question is precisely a community that favours the taking of drugs? This would be, it appears to me like the attempt to treat, without antibiotics (for there is as yet no corresponding therapy for drug dependence) a subject suffering from an infectious disease in an environment full of highly infective germs. I presume that Dr. Smith meant to say "whenever possible, in the community where he lives."

Equally, it appears to me, that it would be well to mitigate the statement that "experience shows that a team comprising side by side with the professionals, non-professionals, ex-drug addicts, is highly efficacious. There is no doubt that the experiment produced interesting results under, at times, difficult conditions for the drug dependents. However, statistics that will prove high efficacy are difficult to obtain. Still, such experiments, as most of those suggested by Dr. Smith, deserve to be undertaken or continued.

DR. GELLHORN: Thank you Professor Jacob. I now call on Dr. Fernand Goffioul who is representing the Academy of Medicine of Belgium and also is a member of the Fédération des bureaux de consultation pour alcooliques et toxicomanes de Belgique. Following Dr. Goffioul we will hear from Dr. Fabian Kamunvi from East Africa and who is representing the East African Medical Research Council of the East African Community.

DR. F. GOFFIOUL (Royal Academy of Medicine of Belgium): Mr. Chairman, I would quite simply like to emphasize a concept which from the practical point of view seemed important to us, i.e. the delimitation of ecological zones of drug addiction and drug dependency. The point of view that I wish to state briefly reflects a synthesis of the experience of practitioners in different disciplines, but particularly psychiatrists practising in hospital out-patient departments and functioning outside hospitals, as well as general practitioners and social workers. The samples on which we have based our opinion are composed of persons who have spontaneously sought advice or who have been subjected to legal guardianship under the terms of various laws for social welfare or regarding suspended sentences. Some patients whom we have taken into consideration are persons undergoing treatment while others are known to us indirectly. What seemed to us essential is that it is not the effects of a drug which it is difficult to treat but the conduct which results in recourse to the drug; the psychological relationship between an individual and a drug is more properly an inter-personal relationship modified by the drug. Among group factors inter-personal influence has seemed to us to be of greatest importance. Indeed, persons are rarely truly isolated: but there is always some contagion and apart from some rare elements of quasi-melancholia it is in general an inter-personal one. When considering group factors allowance must be made not only for the microgroup but also for the macrogroup with its social and cultural features, its doctrinal trends, its shared fashions, but also its shared emotions. At the microgroup level there are real initiatory rites which lead to contagion with the habits characteristic of the group. The examination of one particular case must go into the details of this type of contamination and of the means of influence and persuasion which have been brought into play. At different levels psychological identification, the capacity for reflex internalization of types of conduct and also some slight appreciation of the manner in which the choice of life or the risk of death is assumed and aggressivity is directed to a particular point or remains diffuse, are characteristic parameters which determine the individual roles and the pecking order in these ecological zones. The type of feelings sought for in driving a vehicle is preached explicitly or implicitly, also forms part of the standards of the group and is defined in slogans and stereotyped phrases which are found during clinical questioning to be astonishingly similar from one case to another.

Another fundamental, determinative characteristic is the type of real or phantasmal reaction between doctors, patients and drugs. The drug, being a transitional object, is often imbued with symbolic value. Emphasis must be laid on distortions, falsifications and hedging in the doctor-patient relationship. In this respect, when I was listening to Professor Jacob just now I thought that it is not only in child education that emphasis is necessary but also in faculties of medicine where explicit provision should be made for medical education in this important aspect of the patient-doctor relationship mediated by the medicament. We no longer think that it is geographical or even socio-economic distribution which is applicable in delimiting ecological zones. It is rather not the habitat or the meeting-place but the mode of existence, the mode of co-existence and even, I may say, the mode of remaining solitary in common that must be analysed. Ecological zones can be defined as human groups which, in a way relatively independent of space and time, share at a conscious level and at a socio-cultural level which is partially unconscious a certain way of life and obey addictive urges. At the present time such patients are social deviates and this deviation, with its implications, is an important unfavourable factor in the prognosis. This factor in some cases appears to be more important than the personal relationship to the drug, which varies in frequency, intensity and significance. I should like to describe briefly four types of ecological zone: first the zone of classical drug addictions, notably recourse to opiates and synthetic derivatives and often based on a doctor or medical auxiliary who is himself an addict. These are small groups of not very wide range and their level is stable, constant and endemic. A second type of group or ecological zone is that of drug dependence as a loophole in the doctor-patient relationship, where the prime motive is often to reduce anxiety or to seek to reduce anxiety. A third group, which seems to us important and which is a variant, is that of alcoholics who are unstable. In our experience 30% of alcoholics change their addiction for alcohol to drug dependence and in the case of women 50%. From the point of view of group approach they have special characteristics. The fourth type of group is that of the young people, widely spoken of here, whose ever-changing ideological beliefs and type of conduct involve a far wider range of substances and conduct. They evolve to a much greater extent. From the point of view of public health the greatest social danger threatens persons with drug dependence of type 2, particularly through its effect on road traffic and work places. The untoward social repercussions are considerable. Treatment must be conducted on the basis of affinities and by using group techniques and work in the ecological environment; it must at the same time bring into play the various components of the doctor-patient relationship.

DR. F. KAMUNVI (Secretary, East African Medical Research Council): I thank you very much Mr. President. Before I go to the problem in a developing country, I would like to pose one question to all my senior colleagues here. How would you feel if you were to be the only doctor in a community of over 70,000 people? Not many years ago, I was told by Professor Lambo, who was here with us yesterday, that he was the only psychiatrist in the developing part of Africa. With that question I wish to register our deep heartfelt appreciation for the honour and priviledge that you have extended to us by sending us this invitation to enable us to participate in this importance Conference. Similar sentiments go to WHO and UNESCO. My comments will really be general observations as I am not a specialist; I have drifted from clinical medicine to medical administration.

In listening to the many papers presented and the discussions, I have changed much of my original thinking. I would suggest that the situation in the developing countries is probably much more different today than it was several years ago. In East Africa the problems of alcoholism and drug dependence do actually exist and indeed, it is felt not only among the medical and other workers but also by the public, that it is taking deeper roots. To illustrate this I carried a cutting from one English daily in East Africa which has a good circulation. This excerpt has some disturbing features worthy of note and I would like to read some parts of it. The article is headed:-

"Hippy Glamour Tempts Youth into Drug Danger. Do you have any grass with you brother?" "Grass is the jargon used by young school children when they ask for bhang from people at a popular Nairobi restaurant. This may sound a tall story but you would be shocked to know that quite a number of our school children have been introduced to the habit of bhang smoking. Most of the children come from respected homes and schools in the city. Their parents are working hard and spending large sums of money and trying to make their children good citizens for the future but it looks as if they are wasting their time. Some unscrupulous people who are out to make money clearly are introducing children to the habit which is causing countries like the USA a big headache. The crooks sit at the restaurants ostensibly having a coke but when the children enter - they meet at the back part of the restaurant where bhang and money exchange hands. It is difficult to notice the process since the place is already full of youngsters who gather around a juke box, listening to teenage music. The bhang sellers follow the children even when they are at the proper weekend 'boogie-halls' where they dance during the day. Among the sellers of this dangerous drug are the foreign hippies with shaggy hair and bizaar dress. Apart from showing the children how to smoke bhang, the hippies encourage the kids to behave and dress like them. Already some of our children are using ostensibly vulgar language when talking to their elders and are extremely unpleasant and rude." Then the article goes on describing a few cases of children running out from their parents. One of them was asked whether she liked hippies. "I dig them man, they are the only living people, they are in the 1971, and I'm with them. They are the only people I love." The press asked her would she like some grass? "Oh yeah man! Now you are talking do you have any? I am dying for a puff," she said. She was disappointed however, and extremely rude when she learned that our reporter was joking. Asked whether she was living with her parents, and she said she did not care for them much. The young girl though rude and crude displayed a high degree of intelligence. Unfortunately she is only one example of many others who are landing in the ditch by people calling themselves hippies and members of the permissive society. One parent was interviewed. "I hate the 'boogies' and my daughter would be at home with me and she would now be preparing for University entrance if these places were banned."

A Government Officer from the Child Welfare Society, which in Kenya has been operating for some years, comments in this article that it is the primary duty of every parent to take stern measures to ensure that children are brought up properly. He said that many African parents give their children too much freedom which they do not deserve. Such parents ought to be more strict with their children and bring them up the way they were brought up themselves. I have a big son and daughter and I know what I am talking about.

"What about dance halls where teenagers enjoy the 'boogie' should they continue?" The parents answered "Not at all." Personally I feel they should be banned if these are the places where children are taught how to misbehave in the disguise of recreation. The former chairman of the Kenya Parents Association also thought that the parents are to be blamed for the misfortune that has befallen their children. Some African parents are never at home with their children, they simply give them money to go to the 'boogies' and they, themselves, have a weekend at the bar. This is not good enough. According to a Center on the Abuse of Drugs which was set up in August last year, by the National Council of Women of Kenya, there is a number of school children in Kenya who take drugs, but the problem is not a major one.

Mr. Chairman many of us have followed with responsibility the WHO evaluation of the various Expert Committees on these problems as well as their publications. East Africa is of course trying to learn from these experiences yet the depressing facts of shortage of man power, finance and other facilities cannot indeed make us ready to fight this programme. However, one particular advantage we have in East Africa is that research is recognized and given a high priority and of course WHO will support me on this. In East Africa today there are two fairly distinct communities, the urban and the rural. The latter which form the majority are fairly stable and unsophisticated. The urban however, faces a different situation. Rapid urbanization and industrialization have created a dilemma for us. It is a real social drama. We have thus come with open minds here to seek advice on some of these problems and we hope that the time will come when your Council will meet somewhere in East Africa, so that people will understand that we face a situation which needs your guidance and collaboration. Thank you.

DR GELLHORN: I would think from your description of the situation that you have the dubious pleasure of knowing that East Africa is as developed as many of the other countries about which you have been hearing today. Further, I would think that on the basis of the discussions, Dr. Kamunvi, we have not heard of any specific therapy, but rather a number of possibilities and the street or community center described by Dr. Smith may be as applicable in East Africa, as it is in California.

DR. Y. ROUMAJON (International Society for Criminology): Mr. Chairman, I am a psychiatrist and a specialist in juvenile delinquency. I should like, in the name of the International Society of Criminology, to make some remarks. My first remark concerns a statement made this morning by Dr. Edwards - he said that drug dependence is not a problem of the drug dependent. I find it difficult to agree fully with this statement. This approach has served the study of the external causes of drug dependence. The problem is well known to us since explanation of a very general order and usually neutral, were used to explain juvenile delinquency for the last one hundred years. We have passed the blame from analphabetism to television, and we are now on society. These are explanations that do not prove satisfactory at the level of the individual, considered as a drug dependent.

My second remark refers to the individuality of reactions to the various drugs and to a certain ignorance, I believe, of the real action, at the metabolic level, which some of the drugs may lead to, or may exist side-by-side with, dependence. One example is the case of a patient, followed by me, who suffered from an addiction to sugar, a real dependence with states of need, of anxiety which she satisfied by the massive intake of sugar, piece by piece, reaching up to two or three kilos a day. On the other hand, she was absolutely thin, just as in the case of drug dependents, in the later stages of addiction. There was therefore at the metabolic level, and particularly at the functional level of insulin, a disturbance which could perhaps be observed in relation to other toxic products. To my knowledge, the static chemistry of the cell is well known; its metabolism on dynamic level should be intensely studied, and it is my wish that this should be done.

Finally, a last point, dealing with the relationship between drugs and violence. The problem posed to the criminologist is whether many young people are at present taking drugs to channel a profound violence, which pertains to the human race, and which, whether we like it or not, is restrained by education and by what is known as civilization, but which exists, latent, deep in ourselves. Is the taking of drugs, more or less voluntarily, meant to restrain, or on the contrary does the drug play a role in its release? In a dispensary in the Saint Germain quarter which is the "hot" quarter of Paris in as far as drugs are concerned, Dr. Orsel, the physician in charge noted that whenever the supplies disappeared from the quarter, violence increased and conflicts became more serious all around, with reactions which probably correspond to the electroencephlographic tracings shown to us this morning by Dr. Titeca, where a cortical irritation accompanied privation. It is therefore essential to find out whether the need for violence is appeased by drugs or whether violence is released by them. This is a subject that, in my opinion, could be fruitfully studied. Thank you.

DR. EDWARDS: These are very good questions for debate. I expressed myself clumsily if I am interpreted as saying that it is unimportant to look at the individual. Looking at the individual one of the questions we have to ask is however, "Why is that individual like that?" When we ask that question, we cannot really imagine an individual who has grown up in a state of isolation. An individual has grown up in a society - an individual who will be moulded by that society. To ask how our society treats vulnerable people, whether our society creates vulnerable people, whether our society discards vulnerable people, or whether it supports them, these are therefore legitimate and useful questions, and in no way opposed to the study of the individual.

DR. GELLHORN: There were two questions asked by a member of the panel which have not been answered. I wonder if there is any consensus withregard to registration of drug users?

DR. CAMERON: I would be happy to start a discussion on this. First of all when one speaks about compulsion, it should be borne in mind that it is not an "all" or "none" phenomena. Nor is the source of compulsion singular. Presumably we are talking about legal compulsion as distinguished from other forms. There is very often a great deal of compulsion involved in so-called voluntary treatment that may derive from e.g. family, employer and other pressures. Let us limit ourselves to compulsion imposed as a consequence of legal action. One or several goals and mechanisms might be involved. Provisions might be made for compulsory reporting of cases to assist in epidemiological studies. This could be done without really infringing on the rights of the individual if the reporting were to medical authority without any possibility of making identifying information available to law enforcement officers. This is done, for example in the UK, with respect to the use of certain drugs. A second mechanism might have to do with quarantine, i.e. isolating drug users so they were unable to influence others to take drugs. This is a preventive goal. While the goal is laudable, the mechanism is unlikely to be as effective as one might hope since it would be pretty difficult to get all, or even most, drug users into quarantine. It is unlikely that quarantine *per se* would be very effective if it involved only a few drug users. A third goal of compulsion might be to assure at least some contact with a form of therapy that the person himself had not sought of his own volition. I say contact because it is doubtful that a person can really be compelled to become meaningfully involved in treatment if his own voluntary participation is required. Some contact with available treatment can be compelled, but that is about all. It should also be remembered that compulsion might involve a wide range of possibilities from full deprivation of liberty, the situation persons often think of in connection with compulsion, to as little as the requirement that the person report at stated intervals to a designated person for observation and discussions. The WHO Expert Committee on Drug Dependence, at its last meeting, had the following to say about the effectiveness of compulsory treatment: "The Committee reviewed evidence related to civil commitment for dependence on alcohol and other drugs and reviewed the arguments in favour of and against the use of compulsion in connection with treatment. The Committee considered that the clinical evidence was not sufficient either to support or refute the case for various forms of compulsory treatment, but noted that in spite of considerable experience, compulsory detention alone had not been shown to be beneficial."

One of the major problems associated with compulsion is that it may be used unnecessarily, often out of frustration with the inability to achieve what is desired by other means. It may also be seen as "the answer" and thus stifle the development of other and much more imaginative and potentially useful procedures. If the availability of compulsory treatment were to stifle innovative seeking for new treatment methods, the situation would be counter productive.

I would note finally, that some patients who are compelled to have contact with treatment do not see it as a punitive move. A few will say, "Thank goodness. At last, someone cares enough about me to really try to do something for me. Someone cares." And not a few others who have had enforced contact with good treatment services have benefitted even though their initial reaction to compulsion was negative. But compulsion involving deprivation of liberty, absent the ready availability of good treatment, is imprisonment. Such action should not be taken in the name of treatment.

PROFESSOR JACOB: I would like to ask some preliminary questions. First, I should like to know what is the average duration of the treatment at the moment? Second, what is exactly meant by registration? I think there is a very big difference between anonymous registration and not anonymous registration - and this must be qualified.

DR. GELLHORN: Yes, I think that it is well to define our situation and I would like to interpret Dr. Evang, probably a very hazardous thing. "Is the physician who sees the patient obligated to register the patient either anonymously, in which case it would be for the purpose of gathering statistical data, or by name for the purpose of following the individual patient for therapy or for some other purpose?"

DR. FREEDMAN: I think Dr. Cameron laid out the issues. My own opinion is very much like that of the expert committee, which, I suppose, is a safe opinion. I generally take a cheerful view of the world, and it is only when I get pushed or challenged that I say, "Wait a minute. Let us see what the rush is." This has been my message in the United States. I notice that those people most eager for compulsory treatment and facilities are legislators who very rapidly want to move onto other problems, having solved addiction by passing a law. I have noticed those most eager for registry are either researchers or health officers who are in effect police officers. And I would believe that we would be more authentic if we allowed time to have a real treatment--modalities developed by people who are willing to deliver them. Society had better prove it has some authentic treatment to offer. We would lose very little proceeding this way, leaving aside the issue of registry, which I think in the drug area realistically would accomplish terribly little. Now, as indebted as my profession is to the Scandinavian countries (where elegant record keeping has thrown a light on the etiology of some of the disorders that we have to deal with in psychiatry), I do not think, in the area of drug abuse, that the case finding and definitions are clear enough, or that there have been enough treatment facilities for it to be a wise procedure. And here I meant to answer Professor Jacob on treatment duration; the question is the treatment of what? It is a ridiculous statement these days to be talking about the treatment of a marihuana user, as I have tried to emphasize. Most frequently these people come in with life problems to a variety of agencies--if they are troubled at all. If we are talking about a narcotic addict, we know that, as an addict, he has a problem in life--whatever else may be ailing him. So, it is very difficult to talk about duration of treatment. Generally, when these laws require two years or three, there is simply a kind of mathematical justice related to nothing empirical. So, while I do not think we should put these matters aside, I do not think they have top priority on the agenda. Nobody really discussed this with great passion and seriousness. There are so many problems that have priority. One last thing: If I ever heard anyone pushing a registry, also detail safety for that registry, for its anonymity, and do so with the <u>same</u> kind of passion with which they hate disease and disorder, I would feel a lot more comfortable with registers.

PROFESSOR REXED: I would like to support Dr. Freedman in the case of registers. We had discussed this problem in Sweden in connection with psychiatric diseases generally for research purposes, and some people advocated it. We resisted the idea because we were not sure that this could be done and at the same time safeguarding the individual freedom and right of not being intruded upon. We have not found in the experience of Swedish researchers that there is any difficulty going back to people who have the clinical responsibility of patients if you want more detailed information. So the register that is kept by the individual doctor or researcher or hospital is absolutely enough and you can get into it under the guidance of these doctors for the ordinary statistical information which is anonymous. So my answer as to the need or usefulness of registers would be absolutely "no." There is not special need for them. There can be misuse for them so why discuss them.

I would strongly support what has been said by Dr. Cameron and Dr. Freedman. We have the right to give compulsory treatment in cases of psychiatric disease where there is a danger for the patient or for somebody else or where the patient does not understand that he is in such a dangerous position that he needs treatment. We have not felt that there is a need to extend the right of compulsion to people who have drug problems generally. There we have stopped after very intensive and wide discussion.

We see the treatment going from institutionalized treatment to more open patient treatment and really not treatment anymore but perhaps discussion and psychological handling. Doctors work together with social workers and psychologists in groupings very much like what Roger Smith discussed. If institutionalized treatment is needed, and it may be compulsory if the case falls under the general treatment of law of psychiatric diseases, then we think it should be relatively short and by this we mean that perhaps periods longer than two - three or up to six months would be very inadvisable. We see more and more experimentation in all kinds of guided out-patient situations - in treatment homes - small groups of people under the guidance of medically psychologically or socially trained people. The forms differ according to the inclination and specialization of the people. We have more heartening experience of the results of such treatment, than we have of the ordinary treatment in institutions because there are so many negative factors for the patient associated with the institutions.

We really do not have the doubt expressed by Professor Jacob, that it is necessary or that it is even dangerous to keep the individual in the community while he is being treated, if you call it that. On the contrary we feel that in any case we have to rehabilitate this man or woman socially and then you have to bring him or her back into the family, into his work, into education. There is every possibility if there is a motivation, and we have to create the motivation before it is even possible to start some kind of treatment or rehabilitation. This should be done in the society, in the community. We cannot avoid it anyway.

The comparison with tuberculosis tends to show the weaknesses of the medical model when looking at people with drug problems. In the case of tuberculosis you can hope to clean the environment from all kind of microbes, at least that is what you are trying to do, and in an hygienic society you can do it. In this case of course if the dangers threatening the individual are of quite a different kind and you can never rid society from them because they are part and parcel of the whole setup of our cultural and social environment. So you cannot avoid the community. You will have to have the individual there and I think you will have to try first to develop methods of finding these individuals. As soon as they are motivated, and they often are motivated, although the motivation may oscillate, you will have to try the treatment in the community choosing these different kinds of individual therapy, group therapy, etc. In Sweden we are trying to build up an outpatient system of social centres - medical and social centres combined where we will provide the facilities that society can provide - in medical and social work. These we think should be the centres where in smaller population groups (something like 10 to 15, 000 people perhaps) one could work closely with the people living in the vicinity and also support educational programmes, do all kinds of work with all kinds of people. We look at this from the preventive point of view. Quite a lot of experiments in the line which Roger Smith has described here should be supported.

DR. SMITH: In the United States, compulsive therapy is usually referred to as civil committment and it was regarded, I suppose, several years ago as a fairly progressive alternative to processing through the conventional criminal justice system. I think some very hopeful kinds of experimentations are now beginning to occur in California. For example individuals convicted of drug possession or sales, or drug related crimes, are often given an alternative to incarceration which often involves probation, with a condition that one involves himself in a community based treatment facility. I think that one of the dangers is, that the courts or legal authorities rigidly define what is acceptable treatment. One of the things that I am very anxious about is that, in the case of heroin addiction, the only accepted form of community treatment as an alternative to incarceration will be methadone maintenance. It may be a very simplistic approach to heroin addiction and it is one where we can very carefully control people's movements. I would throw that out as a concern. My further comment is that I am surprised that the whole issue of methadone maintenance has not been raised today. It is a question that is going to become very important in the United States. Perhaps Dr. Freedman can comment on this also. It seems to me that though the feeling in the United States among professionals as well as layman is that methadone maintenance is the answer; that we have finally found the solution. Most of the treatment centres currently being funded include methadone maintenance as a major modality. I am quite surprised that this has not become an issue at this conference.

PROFESSOR BASH: I should like to attempt to reply in a few words to several questions from that particular geographic angle with a representation of which I have been entrusted. Dr. Evang asked: "If you have several hundred thousand peasants cultivating poppies as a cash crop, what does a country do?" As you all know Iran prohibited the cultivation of the poppy in 1956, also the use or the possession of opiates except by prescription. Cultivation was effectively stamped out, the use was not. To reply to Mr. Stewart: Iran was forced to realise that it was a country living among other countries across whose borders these opiates came. Its legislation was changed a couple of years ago. Cultivation under government license is again permitted and also the rationed sale of opiates and other drugs automatically to registered addicts over 60 years of age and on medical prescription to those below 60, thus leading almost to a system of general voluntary registration, which is why I bring up this point. Somebody said yesterday that we do not know what the uses of registration really are. The answer may begin to emerge out of this attempt. Nobody knows how many addicts there are in Iran, for obvious reasons. Informed guess vary between two and four hundred thousand. At the present time about 80 to 85 thousand regular users under the new term of the law have been registered. The registeries are increasing. The form of registration as I mentioned is not compulsory, but brings forward many of those who feel an urgent need for the drug and may thus contribute useful information. Two brief comments to other speakers. I have had considerable opportunity to observe the use of methadone, or rather its misuse, and can only say that I should advise very strongly against it. It has proved in Iran to be totally useless, or rather worse than useless, as it has lulled certain authorities into the comforting but dangerous belief that, in distributing and prescribing methadone rather widely, they are actually doing something to combat addiction. As to the second point regarding compulsory treatment, this also has proved a failure there, though it may work otherwise under other conditions. Thank you.

DR. DENICKER: The question of our President, relating to the compulsory treatment and registration is perhaps somewhat outmoded since in many countries there exists legislation foreseeing voluntary or compulsory treatment in accordance with criteria do not always depend on the initiative of the physician. I believe it is on this point, namely, the need to allow the physician to use his judgement in the choice of treatment. Indeed, there is a true biographic path for the drug dependent who needs at certain times voluntary assistance and at others, as was already said, is content with compulsory treatment. At the moment, and this is a cause for concern, the public authorities and consequently, society, tend to take upon themselves the right of determining under certain conditions, the type of treatment to be administered, basing their judgement not on criteria that are not at all medical, such as recurrences, relapses or even short term recurrences. One would think that in these cases a prolonged compulsory treatment would be more efficacious than a short term treatment. We know however, that the history of the patients does not always conform to this model.

Regarding registration, on the other hand, we have, perhaps, more freedom. I understand that the consensus around this table is not in favour of compulsory registration and compulsory reporting of drug dependents. This may be advantageous in allowing the supply of small quantities of drugs, especially to older persons and to very chronic addicts. But it appears to me that there is one area in which registration, cannot be avoided. I am referring to the field of medical research, to the establishment of an index, which unless it is compulsory for the physician, has absolutely no statistical value. We are fortunate in having in certain countries, medical research institutes with responsible physicians. If in these days, a physician who discovers a drug dependent does not report it to such an institute and if all cases are not registered, no statistics will be of value and the work will be nil.

DR. DUBE: Here we are talking about people who have different needs, who are in different stages of preparedness. We have still a large number of people, probably a lot more than the number we are talking about whose needs are different, whose preparedness is quite different; whose religious and social customs and beliefs are quite different, among whom such procedure that has been mentioned here may not be very easy to apply. For instance, registration. In the first place, facilities may not exist especially when you consider the problem of our 550, 000, 000 people. Even in the big cities where facilities do exist, probably there are not enough trained people who can handle such a problem. And then we must consider their beliefs, the religious, moral and social attitudes. In a country like mine, there are several laws such as compulsory treatment of smallpox, compulsory school attendance, and others which are never applied. A law for the compulsory treatment of drug addicts would remain a dead letter. Treatment facilities are limited. There are only about 300 psychiatrists in the country and in the villages, very few medical men and no facilities.

DR. GELLHORN: I think we have laid that one quietly to rest, if for no other reason, the 550, 000, 000 people. Although I believe there would be a number of people who would set up registers. I can see there would certainly be a difficulty. I wonder if we could very briefly turn our attention to the issue that was raised by Dr. Smith in which he expressed anxiety, and I think the phrase is correct, in regard to the use of methadone maintenance. I interpreted this, not that he was saying that it was no good, but rather that he did not wish to have it as the only therapeutic regimen, but then we had Dr. Bash who said it is a disaster and I would like to have a comment on this apparent disagreement.

DR. FREEDMAN: What I think Dr. Smith is referring to is that methadone maintenance therapy not only has an efficacy, but, in my own opinion, depends on the social group and age group that are most adaptive to it. These happen to be older addicts, I think, generally over 25. They adapt very well to it in our experience and in the experience of a number of others. What Dr. Smith is referring to is that American physicians were mobilized through methadone at long last to come into contact with the addict. It was the simplest way by which to conceive of treating addicts. Now that does not escape the eye of the politician. In fact, the crime rate in Washington, D. C. for certain types of crime dropped. Certain influential persons in that town changed their mind about methadone therapy, and also about what the administration should do about the drug scourge and menace. So, the fact of the matter is that methadone has been sold in the highest quarters as the model, and because it is the most convenient one, it may be extensively pushed as a panacea. I think that is what you are talking about. It is an authentic problem and I do not welcome it. I would hate to see another polarization in treatment approaches as we saw in New York City--a very disastrous episode of polarization as to what treatment is good. I am waiting, as you are to hear Dr. Bash's reservations about methadone. We do see that it can work in some areas, but it is obviously only a start. One last thing: Whatever you think about methadone as a therapy for the range of conditions which have addiction in common, we do have available a great deal of authenticity in treating behavior and character problems. We have to learn how to apply what we know in a variety of different situations in the treatment of people in trouble and alienated people. I think this is happening, whether we think of Dr. Smith's kind of programmes or others. The real problems of professions now is to apply what they know about psychotherapy and psychology--and they are doing it in different situations, whether these are group treatment or retreatist situations, or therapeutic communities. I think this has begun to happen. We simply cannot boast about it yet.

DR. SMITH: Let me begin by saying that, as part of my own treatment programme we have a methadone maintenance component for a very select group of heroin addicts, I find it perhaps, the most effective approach to controllong one of what I consider the three major therapeutic problems. In very general, simplistic terms, you can describe the therapeutic problem as follows: First there is the problem of the drug yen, the pharmacological problem. Secondly is the underlying psychological or psychiatric disorder; that is, the element of compulsion, and finally, there is the question of life style, which is totally exacerbated in the United States by the kind of intense criminal sub-culture that has developed as the result of the way our laws work, as well as the structure of the illicit market place. For the addict that has repeatedly failed in his attempts to detoxify himself; for an addict who has been completely involved in therapeutic programme to no avail; to an addict who seems likely to continue in his addiction, I have found no therapeutic approach that is more effective. It allows the treatment programme to deal with what I consider is very basic therapeutic concerns. Although I have a few physicians on the staff, I think I could probably train a low grade monkey to withdraw someone pharmacologically. It is a very simple problem. But, you detoxify someone and within a few days they have relapsed. With maintained patients, you can get on to dealing with what I consider the most difficult therapeutic concerns;

dealing with the problems of self esteem, the relationship with family, employment - these sort of things. I am not opposed to methadone maintenance. In my own programme we are delighted to have such a component. One of the questions in the United States is, whether or not we can regard methadone for heroin addicts in the same way that one regards insulin for diabetics. There is a school of thought, that says, we must accept the fact that heroin addiction is essentially irreversible and we have to accept the fact then that people are going to be on methadone for the rest of their lives . There is a question of civil liberties involved here that makes me extremely anxious. My own feeling is that once certain aspects of the addicts life have improved, it is our responsibility to encourage weening himself from methadone. Those are basically my concern.

PROFESSOR BASH: I should be glad to comment and to make clear that I entertain no objection nor anxiety about the use of methadone in Dr. Smith's therapeutic programme, also that I quite concede the credibility of his results. I have great anxiety about its use and abuse in so-called developing countries--countries such as Dr. Dube's and those of many of my colleagues, in which a few physicians with scant resources are forced to cope with vast problems as best they can. In some of these countries methadone has been applied as a supposed panacea and has turned into, I assert again, a disaster. The reasons therefore I think Dr. Freedman has very clearly implied, if I may venture to interpret him. He has said that methadone has provided many psychiatrists with their first contact with addicts, but this has also all too often been their last. The treatment usually has consisted, as far as my observations go, in withdrawing the opiate, substituting it by methadone (in other words driving out Satan by Beelzebub), turning the patient loose and telling him to come back for another prescription when the bottle is empty, whereupon the patient swiftly increases his consumption of the less potent narcotic and seldom even returns for a new bottle, but goes back to the familiar opium. Meanwhile the centers and hospitals for the treatment of addicts, convinced that they have done their duty and hampered by an almost total lack of social workers, abandon the difficult task of following up the methadone -substituted addict and prescribe, sometimes under government and even intergovernmental pressure, more methadone, while any realistic treatment or even appraisal of the problem is circumvented. The disaster has occurred. The prime danger in methadone, I think, lies in in its seductive pseudo-simplification, the temptation to use it indiscriminately on a wide scale as a supposed therapy for addiction instead of as an adjuvant to therapy. I do not doubt that it can be useful in a small, selected, carefully supervised, intensively treated group of addicts. I am less optimistic about heroin addicts than Dr. Smith is, but that is another point.

DR. LEONG HON KOON: I think that we very often forget that the person that you are treating is a person, a total person and that any drug be it methadone or anything is not the answer to the treatment. I have often been asked by medical colleagues, through the phone "I have an opium addict, a morphine addict what do you give him Doctor?" Well if the answer was as simple as that he would not have called me up because he would know. If you have an opium addict you give him this, this, that. The simple thing for me to answer is "Well if you really want to give him something to get him out of the withdrawal, give him this prescription, tincture opii, chloral hydrate, give him methadone or give him anything to relieve his pain for the moment. But if you really want to deal with him then talk to him first and see what he wants and what he feels." I think that the most important thing is the first contact with the patient whether you use methadone as your first contact, or whether you invite him from the streets - whether you open a counselling center. The most important thing is contact with the patient, find out what he wants, what he needs and then after that you will get the answers, the things that he will need.

SUMMARY

Dr. Alfred Gellhorn

President of CIOMS

This conference meets the qualification of "an idea whose time has come." As we stated at the outset, there have been many conferences on drug abuse and a number of the participants in this meeting have also been speakers at others, but the mix of many disciplines and the exchange of ideas by experts looking at the general problem from different vantage points has created a stimulating and instructive analysis. Although it is perhaps presumptuous of me to attempt to summarize this round table conference, I should like to indicate some of the impressions which I, as a non-expert will take away. Careful study of the text will provide more instruction.

A noteworthy feature of this conference on drug abuse has been the humanistic approach to the problem by the participants. No punitive or patronizing attitudes were expressed. Rather, concern was shown for the drug users as sick persons--victims of physical and/ or psychic dependence on drugs through the accidents of contemporary living, the design of criminal traffickers, or the deficits of personality associated with psychiatric disorders.

The conference got off to a fast and stimulating start with the reminder that insights into the future could be gained from knowledge of the past. The particularly eloquent admonition which emerged from an historical analysis of earlier ineffective alcohol-abuse control measures was to examine critically the data on which decision making is based.

The abrasive interaction between the younger drug users and the rest of society, the establishment, was clearly recognized and explored. The importance of youthful anxieties and uncertainties was emphasized as a factor in the decision to try drugs. Experimentation can become dependency when drug users are alienated by a condemnatory social attitude. Parents, the health professions, the law and government need to learn their roles in therapeutic programmes if drug abusers are to re-enter the social system.

The drugs most commonly leading to dependence were discussed with regard to their classification, metabolism and mechanism of action as well as many areas of uncertain or absent knowledge. The cardinal and common sense principle that there is a dose-effect relationship was brought out as well as the danger and cruelty involved in the indiscriminate use of the term "drug addict" for occasional users. Although the pharmacological effect of the different groups of drugs is predictable for many of the somatic manifestations, the psychic response of the user was much influenced by the circumstances of the moment. This led members of the conference to speak of the "plasticity" of the agents. Additionally the frequency of the use of combinations of drugs added to the complexity and unpredictability of the psychotropic response. The needs were presented for better methods of detection, for greater understanding of the mechanism of action, and for the development of effective drug antagonists.

As a prelude to the presentations on the prevalence and patterns of drug abuse, there was an informative discussion on the relative importance of the basic personality characteristics of an individual versus the impacts of societal factors in the genesis of drug use, drug abuse, and drug dependence. Although it was quickly accepted that seldom did only one of the triad of Social, Biological, Psychological factors determine the involvement with drugs, the great importance of social factors was repetitively emphasized. The difficulty of recognizing the socio-economic aspects by traditional demographic analysis was demonstrated by drug misuse among the economically poor and also by the poor in purpose among affluent groups. There was discussion of the determinant influence of cultures ranging from greater acceptance by the comtemplative mores of the East to the rebellion of the young against the activist, materialistic culture of the West. The difficulties of interpreting these influences, as well as the effect of many paradoxes in contemporary society such as the varieties of behaviours of law enforcement officers, jurists, school administrators and so on led to a valuable discussion of the need for interdisciplinary research to obtain more meaningful information. Thus the definition of the problem of drug abuse and dependence was recognized to be the behavioural response to the complex interweavings of social, psychological and biological factors.

In a consideration of the prevalence and patterns of drug abuse, the participants reported on the status in France, Scandinavia, Great Britain, Iran, India, Southeast Asia and a part of the United States. It was brought out that in the 1960's drug misuse became pandemic, affecting a progressively younger age group and creating similar patterns of dissemination and use throughout the world.

Control of drug abuse seemed closely connected with the ultimate correction of many of the sociological factors which contribute to anxiety, disillusionment, frustration, desire to escape reality, and rebellion by antisocial behaviour. Additionally however, attention was directed to national legislation, international convenants and education. Although the value of national laws for the control of drugs and drug abuse was questioned, it was noted that there was no information on what might happen in the absence of restrictive laws. Evidence was produced to indicate that the laws are becoming more stringent and punitive against drug traffickers and more lenient on the drug abuser tending to regard him as a sick individual needing treatment rather than as a criminal. This would appear to be an important movement, toward rehabilitation although difficult to implement when the frequency of drug abuse is great and treatment facilities limited.

The international covenants are increasingly effective in limiting the availability of the opiates, but the problems of production, distribution, legitimate and illicit use of synthetic drugs, have yet to be tackled directly by such bodies as the United Nations. Thus prevention of misuse by international restrictions would appear to be a frail reed on which to lean.

The potentional importance as well as the difficulties of educational programmes directed at young people in primary, secondary and higher levels was discussed. As in all educational ventures the effectiveness and credibility of the teacher are crucial and these depend on his real knowledge, his empathy with his audience and their receptivity. The necessity for good education on the "drug problem" in medical schools, law schools, departments of sociology and social work may well prove to be a critical factor both in prevention and treatment.

The discussion of treatment brought out forcefully the complexities introduced by multiple drug usage, the circumstances under which the drug or drugs were self administered, and the types of individuals requiring or requesting treatment ranging from the experimenter, through the occasional user to the physical and/or psychically dependent patient. The success of temporary or long term treatment is also a complex mix of the qualities of the physician or non-physician therapist, the therapeutic environment and the family or community attitudes and resources. For the experimenter or occasional drug user the value of education was related to the trustworthiness of the therapist in the eyes of the patient, the accuracy of his information about drugs and his ability to guide the user to other alternatives for the relief of anxiety or escape or rebellion. Although the recognition of the untoward pharmacological effects of the drugs singly or in combination was acknowledged as important, the frequency with which specific or non-specific antagonists were indicated was considered small. Substitution therapy with methadone for opiate addicts was conceded to have a real place in the management of certain patients, but the popular conception that all those with physical dependence should be on methadone maintenance was dismissed by the experts. Only those with personality defects and demonstrated relapses were considered as candidates for this regimen.

The value of the "street clinics" in America, Scandinavia, and the Far East was presented and this departure from the conventional institutional management appears to be a valuable adaptation to the current needs of the young patients. Community mobilization of resources, community ability through education to modify its attitudes toward drug users so that they are readmitted to society rather than isolated from it, and community capacity to provide alternatives to drug usage constitute critically important aspects of successful therapeutic programmes.

The facts and concepts discussed above were among the highlights of this conference to me, a physician but not an expert. There was a strong consensus among the participants to support the following three conclusions:

1. Research on the non-medical use of dependence producing drugs should be carried out by interdisciplinary groups. The fruitful interchange between epidemiologists, psychiatrists, public health administrators, sociologists, lawyers and pharmacologists at this conference gives the conclusion particular credibility.

2. Education of a wide variety of lay and professional groups to provide understanding of the extraordinary spectrum of non-medical drug usage, the multiplicity and complexity of social, psychological and biological factors in the pathogenesis of the problem and the necessity to facilitate reentry of the patients into the society of the majority rather than keeping them isolated as social deviants.

3. The general medical community must be involved in the management of drug-abuse patients. This is not a therapeutic problem only for experts because there are too few of the latter and because the frequency of drug-dependent patients is so great and their distribution throughout the world is so extensive that management of this problem must be assumed by a large segment of the medical profession.

So I end these impressions of this challenging and rewarding conference. CIOMS expresses its appreciation and thanks to the audience which has been both attentive and contributory to the discussions; to WHO for its hospitality, a special vote of thanks from all of us; and to the participants I can only say that the readers of this round table conference will, I hope, find the book as rewarding as those of us privileged to have heard you during the past two days.

Annex I

THE CONTROL OF DEPENDENCE PRODUCING DRUGS: PRINCIPLES AND TRENDS OF NATIONAL LEGISLATION

Dr. J. de Moerloose

Chief - Health Legislation Unit, World Health Organization

GENERAL OBSERVATIONS:

Until the Second World War, the suppression of illicit traffic in and the abuse of opium, morphine, heroin and a number of other substances having morphine-like properties, as well as cannabis and cocaine, was the primary objective of the narcotics control legislation in the different countries. In 1946 the total number of narcotics subject to international control was still only twenty or thereabouts (most of these narcotics being also subject to control measures instituted by the legislation of the individual countries), but by 1966 this number had increased to 90.[3]

Even prior to the War, legislators and health authorities alike were confronted by two sources of danger to public health. The initial problem was to prevent diversions from the "licit" narcotics market and thus ensure that narcotics were not being employed for non-medical or non-scientific purposes. It proved, however, to be a much more difficult task to combat illicit traffic in narcotics, the considerable dimensions of which have recently been brought to light by a report of the International Narcotics Control Board (1970).[4] Some 1200 tons of opium were reportedly circulating in the illicit market in 1965, compared with the 800 tons which are sufficient to meet world medical requirements. A two-pronged approach must hence be taken by legislative measures and control programmes with the objective of preventing at all costs the spread of drug dependence, a scourge described by President Nixon, in his 17th message to Congress in June 1971, as a "problem which afflicts both the body and the soul of America". It is not exaggeration to declare that the addiction menace has become a veritable nightmare to many Americans. Many other countries are, moreover, now confronted with the problem.

For the legislators of the different countries, the regulatory measures to be taken with respect to the drug addict pose certain problems, the solution to which have often proved disappointing. Moreover, the inherently slow procedure whereby laws are promulgated often results in their being ill-adapted to the situation at a given time; certain substances are thus able to exert devastating effects before being placed under legal control. It is no easy matter for the legislative authority to cope with the rapidly increasing number of hazardous substances. The task is moreover rendered difficult by the expression of discordant views, as for example the propaganda by certain groups in favour of the legalization of cannabis.[6]

An analysis of the legislation of different countries at the present time indicates a very pronounced - and no doubt useful - trend towards the <u>incorporation in a single basic legislative text</u> (or in a group of texts) of the provisions governing the different categories of poisons, including narcotics, stimulants, tranquillizers and sedatives. In France, a Decree dated 19 November 1948 embodying public administrative regulations relating to the importation, possession and use of, and trade in, poisons, divided the various poisons used in medicine into three schedules (in actual fact, this Decree amended an earlier text dating from 1916 which had embodied the same principle). The most stringent provisions of the Decree apply to the substances in Schedule B, in which the narcotics are listed. The supply of these substances to the public is governed by very strict conditions: thus the physician is required to issue prescriptions by the "counterfoil book" (carnet à souches) system, and the dosage must be written out in full. The prescription must be retained by the pharmacist for three years. These are only some of the provisions governing Schedule B substances. Other substances capable of producing dependence appear in Schedules A and C. Transfer from one schedule to another may be effected, depending on necessity. Phenmetrazine, for example, was moved from

Schedule C to Schedule A and then to Schedule B, Barbiturates, on the other hand, appear in Schedule C. It should be mentioned that a certain number of preparations are listed in a special table indicating the exemptions from the provisions applicable to Schedule A, B and C, and based on the pharmaceutical form, composition and, in particular, the concentration of active principle.

Many countries have, of course, adopted such exemption systems and it is a known fact that they may give rise to abuses with serious harmful effects to members of the public. This is true of such substances as codeine, glutethimide, methyprylon, phenobarbital, diazepam and even elixir of paregoric. The consequence of this type of exemption scheme is that these substances, whose ability to produce dependence has been clearly established, are available "over the counter" in many countries. Numerous cases of abusive consumption have been reported.[13]

The importance of the French legislation outlined above is clear, insofar as a large number of French-speaking African and Asian countries have adopted the same legislative approach. As far as English-speaking countries are concerned, the legislative system is often based on a "Dangerous Drugs Act" or "Narcotic Control Act," though the actual provisions vary from country to country. In the United Kingdom, for example, the control of narcotics has been exercised under the Dangerous Drugs Acts and Regulations since 1920. Stimulants, barbiturates and other substances are regulated by other legislation, notably the Pharmacy and Poisons Act 1933 and the Poisons Rules. The tendency to incorporate the various dependence-producing drugs in a single enactment is manifested by the Misuse of Drugs Act 1971 (dated 27 May 1971), which repeals the Dangerous Drugs Acts of 1965 and 1967 as well as the Drugs (Prevention of Misuse Act) 1964. The drugs defined as "controlled drugs" are enumerated in Parts I, II and III of Schedule 2 to the Act. In the United States of America, the Controlled Substances Act (Title II of the Comprehensive Drug Abuse Prevention and Control Act of 1970) contains five schedules of controlled substances, drugs being classified essentially on the basis of their potential for abuse, their accepted medical use and the physical and psychological dependence potential. The United Kingdom approach particularly as regards the provision of narcotics to addicts, is known to have been very strongly influenced by the recommendations of the Rolleston Committee set up in 1924 and, more recently, those contained in the two Brain Reports (1961 and 1965).[8, 10] With the adoption of the Dangerous Drugs Act 1967 and the 1968 Regulations for its implementation, the U.K. legislation became more stringent; heroin and cocaine could no longer be supplied, except under special conditions by general practitioners specifically licensed for the purpose. Provision was also made for the compulsory notification of drug addicts. It is known that certain countries have not adopted, in their legislation, the recommendation that the use of heroin be prohibited.

EXTENSIVENESS AND RAPID OBSOLESCENCE OF THE LEGISLATION

A very considerable number of laws and regulations have been promulgated in the different countries to prevent the abuse of drugs and/or to amend existing texts by providing for the insertion of newly marketed substances, some of which may constitute a serious danger to public health. The legislation has become so voluminous that only a broad outline of the current situation can be given in this paper.

In the United States of America, for example, the Comprehensive Drug Abuse Prevention and Control Act of 1970 was intended to consolidate some 50 public laws on the subject.[1] Regulations for the implementation of the 1970 Act were issued on 20 April 1971. While this updating of the legislation does clarify the position, at least at the Federal level, it is obvious that the new Act must be interpreted in the light of other Federal Acts; moreover, the different States have promulgated hundreds of enactments having the same objective. In addition, the extensive jurisprudence has to be taken into account in analysing these various items of legislation.

Amendments to the legislation may be frequent. Information supplied to WHO some two years ago concerning the regulation of substances not included in the Single Convention of 1961, and in particular hallucinogens (which were included in Schedule H to the Food and Drugs Act of Canada), is already out of date. The present situation in Canada is that hallucinogens, commercial transactions of which are prohibited, are included in a new Annex J (Restricted Drugs); their possession is subject to very strict rules and severe penalties are prescribed for possession for the purpose of trafficking and for trafficking itself. Narcotics as such are controlled in Canada by the Narcotic Control Act. Amphetamines and barbiturates appear in Schedule G to the Food and Drugs Act, and, as far as possession, the penalties for trafficking etc., are concerned, are subject to provisions different from those applicable to narcotics. The different Canadian provinces have their own legislation on such matters as the treatment of addicts, the practice of pharmacy, etc. The same pattern recurs in other countries having a federal-type structure such as Australia and Switzerland, where narcotics and other drugs capable of producing dependence are governed by federal provisions and by the legislation of the different jurisdictions constituting the Federal State. In the case of Switzerland, narcotics are covered by the Federal Law of 1951 on narcotics, whereas amphetamines are regulated by the different cantonal laws. The latter follow the recommendations of the Office intercantonal de Contrôle des Médicaments (OICM),[5] which classifies pharmaceutical substances into five groups; both of the first two groups require a medical prescription, non-refillable in the case of Group B.

DESCRIPTIONS OF ONE TYPE OF NATIONAL LEGISLATION

The Norwegian legislation is described below by way of illustration. The most important Norwegian provisions dealing with narcotics are contained in Part VI of the Law of 20 June 1964 concerning medicaments and poisons, etc. Section 21 prescribes that authorization is required for the manufacture, import, export and carriage in transit of narcotics. The circumstances in which narcotics may be used, for medical and scientific purposes only, are laid down in the Regulations concerning narcotics, made in pursuance of the Law. Severe penalties are imposed for the illicit possession of narcotics. Under a recent amendment to the Penal Code (Section 162), persons engaging in trafficking are now liable to up to six years' imprisonment. Under Section 23 of the above-mentioned Law, any person may, regardless of statutory professional secrecy, inform the Director-General of Health Services of all confirmed or suspected cases of abuse of dependence-producing drugs. Members of the medical and pharmaceutical professions, the police, and temperance boards must, if requested, inform the Director-General of all confirmed or suspected cases of narcotics abuse. The Director-General, in turn, supplies the above-mentioned categories of persons with any pertinent information he possesses in this field. The provisions governing the import, export and transit of narcotics, as well as the conditions under which they may be held in aircraft and ships, are prescribed in detail by the Regulations of 6 January 1965, which also lay down requirements in regard to the recording of commercial transactions and specify that while the opium poppy may be cultivated under a special licence, the cultivation of cannabis and coca is prohibited. The substances listed in Annex I are in fact those appearing in the corresponding Schedule of the Single Convention. It is interesting to note that amphetamines and hallucinogens were incorporated in Annex II (substances whose manufacture, import, export, sale, possession, use, etc. are prohibited) to the Regulations by an amendment dated 26 March 1971. Under the terms of the amendment (which became effective on 1 July 1971), amphetamines may however be supplied for medical use under a special licence. The control of other substances liable to produce dependence, i.e. analgesics, tranquillizers and barbiturates, is governed by a Crown Resolution of 19 February 1965, which specified that these substances may be procured only on medical prescription, the latter being refillable only when the physician has expressly stated this in the prescription. These substances appear in Annex I (Class C) to the Crown Resolution. Insofar as hallucinogens have not been accorded registration as medicaments, their sale is prohibited. The Instructions of the Director-General of Health Services on the prescription of substances which may produce addiction (issued in April 1958 and amended in 1958, 1962 and 1965) are intended to guide physicians on the use of narcotic drugs in practice. The different drugs are divided into a number of groups according to therapeutic use and dependence liability. The Instructions indicate that precautions to be taken in prescribing the various drugs to four categories of patients, viz. (a) persons who are not regular patients of the physician, (b) regular, non-addicted patients, (c) regular patients who are both chronically ill and addicted to drugs, and (d) regular patients assumed to be suffering from drug addiction alone. In the latter case, it is recommended that narcotic drugs be prescribed only if the patient agrees to be hospitalized as soon as such treatment is made available and in the meantime follows the physician's instructions.

INCREASING STRINGENCY OF LEGISLATION

There has been a general trend in recent years towards the promulgation of stringent provisions concerning the use of narcotics for non-medical purposes and, in particular, to combat trafficking in these substances. As has already been noted in the case of Norway, certain countries have even amended their Penal Code by the introduction of more severe penalties for offenders. In France and the United Kingdom, recently adopted legislation has been aimed at the rigorous suppression of illicit traffic in narcotics; certain provisions concerning drug addicts are included in these texts. In France, for example, compulsory disintoxication treatment for illicit users of narcotics is prescribed by a Decree dated 19 August 1971. This Decree was promulgated in pursuance of the Law of 31 December 1970, which contains drastic provisions aimed at the suppression of trafficking in narcotics. The now generally accepted principle that a clear distinction must be made between the (drug-dependent) patient who contravenes the legislation and the dope-peddler whose sole aim is ill-gotten profit is among those applied by the Law in question. It lays down particularly severe measures against dope-peddlers, especially repeated offenders. Facilitating the use of narcotics by persons less than 21 years of age is punishable by five to ten years' imprisonment; in addition, the court may deprive the offender of his civil rights for a period of the same duration. Restrictions on residence and withdrawal of the offender's passport are other possible penalties for trafficking, provided for by the Law.

In the United Kingdom, the Misuse of Drugs Act 1971 imposes more stringent provisions than an earlier Act of 1964. The mere unlawful possession of a "controlled drug" is regarded as a serious offence. The Act is however primarily designed to combat drug trafficking, the severity of the penal provisions varying according to the nature of the illicitly distributed substance. The maximum term of imprisonment is now 14 years, compared with the 10 years prescribed by the Drugs (Prevention of Misuse) Act 1964. A medical practitioner who infringes the provisions of the Law may be forbidden to prescribe, supply or possess specified controlled drugs.

In Israel, a term of ten years' imprisonment is imposed, under the Dangerous Drugs Ordinance (Amendment No. 2) Law, 1968, on any person instigating a minor to use "dangerous drugs". The following acts are specifically mentioned: (1) giving a dangerous drug to a minor: (2) being a person responsible for a minor, allowing the latter to obtain or use a dangerous drug; (3) instigating a minor to obtain or use a dangerous drug.

Although a number of legislative texts adopted in Iran in 1969 provided for the re-introduction of limited (and controlled) poppy cultivation, it was also prescribed that drug-dependent civil servants and workers failing to give up their addiction within the specified period are liable to dismissal or suspension from duties, with expulsion from their institutions being the penalty in the case of students. Moreover, under the terms of the Law of 10 June 1969, the death penalty is imposed on any person illicitly importing, selling, offering for sale or holding more than 2 kg of opium or more than 10 g of morphine, heroin or cocaine. Where the quantities involved are lower, the penalty is three to fifteen years' imprisonment with hard labour.

A recent amendment to the Polish Penal Code (Section 161) prescribes that any person who, not being authorized to do so, supplies a narcotic to another person or incites him to take a narcotic is liable to up to five years' deprivation of liberty.

Severity towards physicians or pharmacists responsible for drug addiction of "iatrogenic origin" is another trend manifested in the legislation of several countries. Diverse measures, including the withdrawal of the right to practise, may be taken against members of these professions. In Argentina, for example, contraventions of the provisions of the Law of 29 July 1968 on narcotics may entail the withdrawal of the right to practise for a period of up to three years. Abuse prescribing of narcotics is considered to be a lapse of discipline in France. Article 629 of the Public Health Code had laid down that a physician guilty thereof could be prohibited from practising for a two-year period while the 1970 Law made this penalty more severe by extending the period of prohibition to up to five years. In Sweden, the Disciplinary Council of the National Medical Board may, in the case of a physician guilty of the abuse of narcotics or the unlawful prescribing thereof, caution or warn the physician, revoke his right to prescribe narcotic drugs or limit this right to a single pharmacy, or even revoke his right to practise medicine.[7] In Switzerland, Section 12 of the Federal Law of 1951 on narcotics empowers the cantons to deprive physicians, dentists, and the managers of pharmacies serving the public, who become addicted to narcotics or abuse the privileges accorded them by the legislation, of the right to prescribe narcotics, have them in their possession, etc: such deprivation may be temporary or definitive in character.

In Norway, special measures to prevent the abusive prescribing of narcotics had already been prescribed by a Law of 28 June 1957. Under the terms of this Law, the Ministry of Social Affairs may issue a warning to a physician whom it considers is making abusive use of prescriptions for narcotics. If the physician ignores the warning and refused to relinquish voluntarily his right to prescribe narcotic drugs, the Ministry may inform a special Council for the Prescription of Narcotics; this Council has the power to withdraw his right to prescribe narcotics for a period of up to five years.

TREATMENT OF DRUG-DEPENDENT PERSONS

While the basic objective of the type of legislative text outlined above is to control illicit traffic in narcotics, certain provisions concerning drug-dependent persons are occasionally encountered. This is the case, for example, in Tunisia, where Title VII of the Law of 26 July 1969 prescribing regulations con-concerning poisons deals with the detection and treatment of drug-dependent persons. However, legislation dealing specifically with the latter problem and extending to alcoholics as well is more and more frequently encountered. This trend has been evident in Australia, Canada, Finland, Norway and New Zealand, to cite some examples. Another tendency is to cover alcoholics and addicts alike in the general legislation on mental patients. In Romania, for example, a 1965 Decree dealing with dangerous mental patients applies by extention to dangerous addicts and alcoholics. This approach has proved useful in practice since the procedure governing the hospitalization of addicts is based on the same principles as that applicable to mental patients, the systems of voluntary or compulsory admission to treatment centres being applicable in both cases.

Another trend in the legislation is to make compulsory the notification of confirmed or suspected cases of drug dependence to the health authority. A parallel has been drawn in this context-with the notification of communicable diseases. In a survey which we published in 1962 on the then existing legislation on the treatment of drug addicts, we drew attention to the notification requirement. At the time, compulsory notification procedures were already in force in Argentina, Brazil, Ecuador, Italy, Panama, Venezuela, Western Australia, etc.

Notification was made compulsory in Japan in 1950, a central register of drug addicts being kept by the Ministry of Health and Welfare.[9] A Spanish Law of 1967 prescribes that all confirmed or suspected cases of drug addiction must be reported to the competent Provincial Health Office. In Switzerland, the Federal Law on narcotics authorizes physicians and pharmacists to notify any cases of drug addiction of which they may become aware in the course of their professional practice. In Tunisia, physicians are required to notify the National Bureau of Narcotics (Bureau national de Stupéfiants), under the terms of the Law of 26 July 1969. In New Zealand, any Medical Officer of Health who considers that a person is or is likely to become an addict is empowered to furnish appropriate information to physicians and pharmacists with the objective of preventing the illicit acquisition of narcotics by the person. On the subject of notification, it may be mentioned that, insofar as the legislation of most countries requires that records be kept of the names and addresses of persons for whom prescriptions for narcotics are dispensed, the inspectorate of pharmacies or its equivalent is able to detect cases of abusive prescribing and hence identify drug addicts.

We present below some examples of legislation on the treatment of drug-dependent persons, notably for the following countries: France, Australia (Victoria), Norway, the United States and Romania.

The French Law of 31 December 1970, which amends the Public Health Code, establishes the general principle that any person making illicit use of narcotics must be placed under the surveillance of the health authority. Two forms of treatment of addicts are provided for, depending on the results of a medical examination, viz. (1) disintoxication treatment in an approved establishment, or (2) medical surveillance by a social hygiene clinic or an approved public or private health establishment. There are four different procedures whereby persons may be subjected to disintoxication treatment or medical surveillance, as follows:-

1) on the initiative of the <u>public prosecutor</u>, who may order persons who have made illicit use of narcotics to submit to the above-mentioned measures (the health authority is informed and arranges for the necessary medical examination and investigation): first offenders who have complied with the medical treatment prescribed for them and have continued the treatment until its termination are not liable to penal proceedings;

2) on the initiative of the health authority (a procedure which may be regarded as "administrative" rather than judicial), when it is informed, either by the certificate of a physician or the report of a social worker, that a person is making illicit use of narcotics;

3) on the voluntary attendance of the addict at a clinic or hospital for the purposes of treatment, in which case the person is not subject to the control of the health authority and may have his identity kept secret if he so wishes;

4) by writ of the examining magistrate of the juvenile court magistrate, or by court order confirming the writ or extending its validity, in the case of persons charged with the offence of making illicit use of narcotics and for whom medical treatment is found to be necessary; such persons are enjoined to undergo disintoxication treatment accompanied by all medical surveillance and rehabilitation measures appropriate to their condition. The detailed procedures governing this system of commitment were covered by a Decree dated 19 August 1971.

It is clear from the above that, as in various other countries, drug-dependent persons are no longer treated as criminals even if they contravene the legislative provisions in force, but primarily as patients, an approach which a WHO Expert Committee had recommended in 1966.[14]

In Australia (Victoria), the Alcoholics and Drug-dependent Persons Act 1968 repealed the Inebriates Act 1958. The 1968 Act provides for voluntary admission to a residential treatment centre on condition that the drug-dependent person undertakes to remain in the centre; if necessary, he may be held for a period of three months after admission. Involuntary commitment is made either by court order or on the basis of a complaint lodged with the judicial authorities by a relative, a member of the police, or a welfare officer. After the statutory period of observation, the medical officer in charge of the assessment centre may commit the person to a treatment centre, provided that two medical practitioners have certified that the person is an addict. The Law empowers the courts to order the release on recognizance of a person sentenced to a term of imprisonment for any offence in the commission of which drug dependence was a contributory factor, on condition that the person seeks treatment in a treatment centre, either as an in-patient or out-patient, for a period of not less than six months and not more than two years. Any person who conforms strictly to the conditions of his recognizance during the prescribed period is discharged from his original sentence. The Inspector of Treatment Centre may, if he thinks fit, recommend to the court the early discharge from the treatment centre of a person released on recognizance.

In Norway, the temperance boards created by a Law dated 26 February 1932 were assigned responsibility for measures concerning drug addicts by a Law of 6 July 1957 (amended in 1964). Temperance boards operate in each commune. There is a similar system in Finland, where social welfare boards have been created for the purpose. Under the Norwegian legislation, the treatment or hospitalization of drug addicts may be affected on a voluntary basis or by persuasion, or on a compulsory basis. In the latter case, the period of hospitalization may not exceed two years. Requests to the temperance board for the commitment of a person may be submitted by the latter's family, by a physician or by other competent parties. If the board orders an addict to enter hospital and he fails to comply, the board may decide that he be committed for a period of up to two years in cases where the person: (1) ill-treats his spouse or children or exposes his children to moral danger or neglect; (2) fails to fulfil his maintenance obligations under the law: (3) exposes himself to bodily or mental harm, endangers his own life or that of others, or repeatedly makes himself a public nuisance; (4) becomes a burden on public assistance or on his family; (5) dissipates or gambles away his worldly goods to such an extent that he or his dependants may become indigent.

The board's proceedings in such cases are presided over by the district or local magistrate. The latter may either reject the request out of hand or issue a detention order. In the case of voluntary admission, the addict may be detained in the establishment (assuming this is necessary to assure his recovery) for the period agreed upon by him at the time of his admission, provided that such period does not exceed two years.

As far as the United States is concerned, it is a known fact that the provisions of the Harrison Act of 1914 were interpreted as prohibiting physicians from administering narcotics to addicts. This system of "treating" the addict has been contrasted with what is often called the "British approach" to treatment, which enabled physicians, under certain conditions, to administer narcotics to addicts undergoing ambulatory treatment. Drug addiction was, moreover, long regarded in the United States as a criminal act liable to punishment. The most significant changes in this field were introduced by California legislation in 1961 and by New York State in 1962 (the so-called Metcalf-Volker Law).[2] The New York law provided for the following forms of commitment of drug addicts to appropriate facilities: (1) voluntary admission for a period not exceeding 30 days (subject, however to extension under certain conditions); (2) admission for a period not exceeding 12 months on court certification, based on the written certificate of two physicians (petitions for admission could be presented by a relative of or person residing with the addict or by the addict himself); (3) a defendant arrested on a criminal charge and detained in custody, who requested or consented to a medical examination to determine whether he was addicted, was (if found to be an addict) subjected to a medically supervised detoxification treatment, after which the court could decide whether he was to be civilly committed as an addict; (4) a defendant arrested on certain narcotics charges could request to be civilly committed to a hospital facility as a narcotic addict, in which case he was required to undergo a medical examination and (if found to be an addict) submit to mandatory civil commitment and aftercare supervision, it being understood that if he successfully completed this medically supervised in-patient and aftercare programme, the criminal charges pending against him would abate.

The Metcalf-Volker Law was considerably amended in 1966 by the so-called Rockefeller Law. This created a Narcotic Addiction Control Commission and provides for two sets of procedures applicable to narcotic addicts, viz: (1) in the case of non-delinquent addicts, a justice of the Supreme Court or a judge of a county court may certify an addict to the care and custody of the Commission, upon application by a person residing with the addict, a relative, anyone who believes that a person is an addict, or a person believing himself to be an addict (if the results of a medical examination are positive, the person may be committed for up to 36 months); (2) in the case of addicts (or suspected addicts) arrested on certain criminal charges, a medical examination is performed and, if the person is found to be an addict and pleads guilty or is found guilty, one of the following procedures is adopted:

(a) where sentence is to be imposed for a misdemeanour or for the offence of prostitution, the defendant is certified to the care and custody of the Commission for a period to terminate on the discharge of the defendant by the Commission as being rehabilitated or on the expiration of 36 months, whichever is first:

(b) where sentence is to be imposed for a felony, the court may either impose the penalty applicable in accordance with the Penal Law or certify the defendant to the care and custody of the Commission for a period to terminate on the discharge of the defendant by the Commission as being rehabilitated or on the expiration of 60 months, whichever is first.

As far as youthful offenders are concerned, the provisions of paragraph (a) above are applicable if the offender is adjudicated to be a narcotic addict.

At the Federal level, the Narcotic Addict Rehabilitation Act was passed in 1966.[16] This Act made provision for the possibility of civil commitment for drug addicts who have been charged with or convicted of violating Federal criminal laws or who request treatment. Thus, an "eligible individual" (charged with certain offences against the United States) who elects to be civilly committed in lieu of prosecution is (if determined to be an addict) committed to the Surgeon General by the court for examination, treatment and rehabilitation. If the treatment is successfully completed, the charge is dismissed. Title II of the Act deals with "eligible offenders". Addicts in this category are committed to the custody of the Attorney General for a period not exceeding ten years. Title III deals with the civil commitment of persons not charged with any criminal offence. In such cases, the Federal district court may commit the addict to the care and custody of the Surgeon General for treatment. A Federal district court may also order commitment if a petition has been lodged, not by the addict himself but by a "related individual". Title IV contains provisions dealing with rehabilitation and post-hospitalization care programmes for discharged addicts.

It appears that, in Japan, rigorous measures for compulsory hospitalization have succeeded in limiting the advance of the scourge of addiction, this being the case for amphetamines as well as heroin. It has been reported, for example, that some 900 mental hospitals were designated as treatment centres for heroin addiction in 1963.[9]

The Romanian Decree of 25 January 1965 applies not only to dangerous mental patients and alcoholics but also to drug-dependent persons considered as dangerous because they endanger their own life, health or bodily integrity or that of others, and are liable to commit other serious acts covered by the penal legislation, or because they repeatedly and seriously disturb, by their behaviour, the normal living or working conditions of other persons. In addition to being subject to such compulsory medical treatment as may be prescribed, they may also be committed to an institution for medical treatment. This procedure is affected in two stages. In the initial stage, the public prosecutor is responsible for taking the decision on the "provisional hospitalization" of the patient in the most appropriate specialized medical establishment. In the second stage, i.e. after the addict has been examined by the medical board of the establishment, a decision that the person be detained for medical treatment may be taken, the consent of the competent people's court being required.

According to the WHO Expert Committee on Drug Dependence[17] compulsion could be used: (1) to provide care or treatment that the individual does not desire to a form of treatment other than that which he prefers: (2) to invoke the principle of qua rantine by regarding the individual as a carrier of a communicable disorder that seriously threatens the health of the community. In the Committee's view, however, "compulsory detention alone had not been shown to be beneficial."

CONCLUSIONS:

The control of the substances capable of producing dependence, with a view to preventing their utilization for non-medical purpose, is governed by what has become a body of law of considerable magnitude. The Regular publication by the United Nations of the laws and regulations promulgated to give effect to the provisions of the international treaties on narcotic drugs (E/NL series) enables an up-to-date picture to be obtained of legislation in force. Many other items of legislation are however not covered by the United Nations publication. An awareness of this legislation is essential in order to understand the overall framework of national pharmaceutical legislation within which the narcotics legislation is drawn up. This is, for example, the case for the regulations dealing with the trade in and distribution of poisons, the treatment of drug-dependent persons, and the measures applicable to diverse substances not covered by international treaties. This lacuna is filled by the publication, by WHO, of the International Digest of Health Legislation, which accords a prominent place to the abovementioned areas of legislation. In 1962, for example, a survey of legislation on the treatment of drug addicts was published.[15] As far as other publications are concerned, those of the Council of Europe dealing with the narcotics legislation of its Member States should be noted.

Knowledge of the legislation of different countries may enable amendments to be made to national legislation and also helps in identifying gaps in the latter. With regard to Sweden, for example, Goldberg[7] noted in 1968 that "stimulant drugs... are not considered narcotic drugs, and often are not even on prescription, in many of the countries where they can be bought. This means that in many instances Swedish authorities have met with very little co-operation from authorities in other countries, when trying to stop illicit trade by investigating a transaction deemed as criminal in Sweden, but which is considered legal in the country where the deal was made".

The study of national legislation is of very considerable interest since there is no doubt that international conventions cannot be drawn up without regard to the prevalent national legislative measures. A striking example is afforded by the absence, in certain countries, of a ban on heroin, which influenced the 1961 Single Convention.

We have given an outline of both French and British legislation in this paper on account of the fact that many countries have taken one or other as a model for their own legislation. Our account of the Norwegian legislation was broader in scope, so as to enable a picture to be obtained of the measures adopted in one of the countries of Northern Europe for which comprehensive information is available. The paper also includes an analysis of the most important items of legislation in several other countries.

In general, a greater stringency can be discerned in the provisions dealing with the illicit use of drugs, the distinction between offences committed by the addict and by the dope-peddler being increasingly pronounced. In some countries, this stringency has been manifested by amendments to the penal code, while elsewhere, as in France and the United Kingdom, special legislation on the subject has recently been adopted. The change of attitude towards the problem has also been reflected in the treatment of addicts. Several countries now enable the delinquent addict to undergo disintoxication treatment as an alternative to imprisonment. Successful completion of the treatment may lead to the suspension of penal proceedings. A number of countries have established a central narcotics bureau within the health ministry, with special responsibility for the implementation of the legal provisions on dependence-producing substances. The corresponding bureau in the United States was transferred to the Department of Justice in 1968.

In regard to alcohol, it must be recognized that few legislative measures have been taken to restrict its consumption and that, quite to the contrary, massive advertising is carried out to encourage consumption. On the other hand, certain countries have placed the treatment of alcoholics on the same footing as that of drug addicts.

The importance of national legislation and the obligations of States under the Single Convention on Narcotic Drugs, 1961, was the subject of a paper published by Reuter in the Bulletin on Narcotics (1968).[11] As a corollary, Vaille later presented in the same publication (1969) a Model Law for the application of the Single Convention on Narcotic Drugs.[12]

It should be stressed that, in the control of drug misuse, legislation not supported by educational, social and/or economic measures as well as comprehensive treatment and rehabilitation of drug-dependent persons, is bound to failure.

REFERENCES

1. BNDD Bull., 3, No. 2, p. 6. 1971
2. Brill, H. (1970), Drug Addiction and dependence in New York State: a programme for prevention and control. In: Phillipson, R.V., ed., Modern trends in drug dependence and alcoholism, London, Butterworths, pp. 3-21

3. Bull. Narcot., 1966, 18, No. 1, p. 1
4. Bull. Narcot., 1970, 22, No. 3, p. 39
5. Campanini, N. (1967) Bull. Narcot., 19, No. 2, p. 13
6. Canada, Ministry of National Health and Welfare (1970) Interim Report of Commission of Inquiry Into the Non-Medical Use of Drugs, Ottawa
7. Goldberg, L. (1968) Bull. Narcot., 20, No. 1, p. 1: No. 2, p. 9
8. Jeffery, C. G. (1970) Drug control in the United Kingdom. In: Phillipson, R. V., ed., Modern trends in drug dependence and alcoholism, London, Butterworths, pp. 60-74
9. Nagahama, M. (1968) Bull. Narcot., 20, No. 3, p. 19
10. Phillipson, R. V. (1970) The implementation of the Second Report of the Interdepartmental Committee on Drug Addiction. In: Phillipson, R. V., ed., Modern trends in drug dependence and alcoholism, London, Butterworths, pp. 75-98
11. Reuter, P. (1968) Bull. Narcot., 20, No. 4, p. 3
12. Vaille, C. (1969) Bull. Narcot., 21, No. 2, p. 1
13. Varenne, G. (1971) L'abus des drogues, Bruxelles, Dessart
14. WHO Expert Committee on Mental Health (1967) Services for the prevention and treatment of dependence on alcohol and other drugs (Wld Hlth Org. techn. Rep. Ser., No. 363)
15. World Health Organization (1962) Treatment of drug addicts. In: Int. Dig. Hlth Leg., 1962, 13, 3
16. Yolles, S.F. (1970) The Narcotic Addict Rehabilitation Act of 1966 and its implications. In: Phillipson, R. V., ed., Modern trends in drug dependence and alcoholism, London, Butterworths, pp. 99-115
17. WHO Expert Committee on Drug Dependence (1970) Eighteenth Report, Wld Hlth Org. techn. Rep. Ser., No. 460

LEGISLATION

Argentina: Decree No. 3450 of 1944 making the notification of habitual addiction to narcotic drugs compulsory, incorporated in Law No. 12912 published on 19 December 1946. Law No. 17818 of 29 July 1968 (see E/NL. 1968/48)

Australia (Victoria): The Poisons Act 1962 (see Int. Dig. Hlth Leg., 1965, 16, 17)

The Alcoholics and Drug-Dependent Persons Act 1968 (see ibid., 1970, 21, 244)

The Drugs of Addiction and Restricted Substances Regulations 1968 (No. 2) (see E/NL. 1969/60)

Canada: The Food and Drugs Act and Regulations (1971 consolidation)
The Narcotic Control Act
Finland: Law No. 96 of 10 February 1961 on the treatment of persons making improper use of intoxicants (see Int. Dig. Hlth Leg., 1962, 13, 249
France: Decree of 19 November 1948 embodying public administrative regulations relating to the importation, possession and use of, and trade in, poisons
Law No. 70-1320 of 31 December 1970 relating to the health measures for the control of drug dependence and the suppression of traffic in, and illicit use of, poisons (see Int. Dig. Hlth Leg., 1971 22, 233) *
Decree No. 71-690 of 19 August 1971 establishing the conditions under which persons who have made illicit use of narcotics and are charged with a contravention of Article L. 28 of the Public Health Code may be compelled to undergo disintoxication treatment
Iran: Law of 4 March 1969 permitting limited cultivation of poppy and export of opium (see E/NL. 1970/17)
Executive Code under Articles 4, 5, 7 and 7 of the Law of 4 March 1969 permitting limited cultivation of poppy and export of opium (see E/NL. 1970/20)
Law of 10 June 1969 intensifying punishment of main perpetrators of the crimes mentioned in the amending Law of the Law banning poppy cultivation and permitting abolition of the prosecution and punishment of other perpetrators of the said crimes (see E/NL. 1970/24)
Israel: The Dangerous Drugs Ordinance (Amendment No. 2) Law, 1968 (see Int. Dig. Hlth Leg., 1971, 22, 319)
New Zealand: The Narcotics Act 1965 (see Int. Dig. Hlth Leg., 1967, 18, 682 and E/NL. 1967/5)
The Narcotics Regulations 1966 (see E/NL. 1967/6)
The Alcoholism and Drug Addiction Act 1966 (see Int. Dig. Hlth Leg., 1968, 19. 421)
The Alcoholism and Drug Addiction (Forms) Regulations 1968 (see ibid., 1970, 21, 776)
Norway: Law No. 8 of 28 June, 1957 to amend the Law of 29 April 1927 on the rights and obligations of physicians (prescriptions for medicaments containing narcotics)
Law No. 1 of 6 July 1957 to amend the Law of 26 February 1932 on temperance boards and the treatment of alcoholics

* Comprises only those items of legislation actually analysed in the paper.

Instructions on the prescription of substances which may produce addiction. Drawn up by the Director-General of Health Services in April 1958 (amended in 1958, 1962, and 1966) (see E/NL. 1966/53)

Law No. 15 of 5 June 1964 to amend the Law on temperance boards and the treatment of alcoholics and drug addicts (see Int. Dig. Hlth Leg., 1968, 19, 658)

Law No. 5 of 20 June 1964 concerning medicaments and poisons etc. (see Int. Dig. Hlth Leg., 1965, 16, 154)

Crown Resolution of 6 January 1965 embodying regulations concerning narcotics, etc. (see E/NL. 1966/52)

Crown Resolution, 19 February 1965 embodying regulations for the ordering of pharmaceutical products and their supply to pharmacies (see Int. Dig. Hlth Leg., 1966, 17, 124)

Law of 14 June 1968 to modify the penalties imposed for contraventions of the laws and regulations on narcotics.

Poland: Law of 19 April 1969 to amend the Penal Code (see E/NL. 1970/28)

Romania: Decree No. 12 of 25 January 1965 concerning the medical care of dangerous mental patients (see Int. Dig. Hlth Leg., 1966, 17, 140)

Spain: Law No. 17 of 8 April 1967 to bring up to date the legislation in force on narcotic drugs and make it conform with the provisions of the 1961 Convention (see E/NL 1967/43)

Switzerland: Federal Law of 3 October 1951 on narcotics, as amended on 1 May 1953

Tunisia: Law No. 69-54 of 26 July prescribing regulations concerning poisons (see Int. Dig. Hlth Leg., 1970, 21, 660)

United Kingdom: The Pharmacy and Poisons Act 1933, as amended

The Dangerous Drugs Act 1965 (see E/NL. 1965/24)

The Dangerous Drugs Act 1967 (see E/NL, 1969/30)

The Dangerous Drugs (Notification of Addicts) Regulations 1968 (see Int. Dig. Hlth Leg., 1969, 20, 805)

The Dangerous Drugs (Supply to Addicts) Regulations 1968 (see ibid., 807)

The Poisons Rules 1970

The Poisons List Order 1970

The Misuse of Drugs Act 1971

United States of America: The Comprehensive Drug Abuse Prevention and Control Act of 1970

Regulations implementing the Comprehensive Drug Abuse Prevention and Control Act of 1970. Dated 20 April 1971

New York: An Act to amend the mental hygiene law, in relation to drug addiction, and repealing section three thousand three hundred forty-one of the public health law, relating thereto. Dated 21 March 1962. (The "Metcalf-Volker Law") (see Int. Dig. Hlth Leg., 1964, 15 897)
An Act to amend the mental hygiene Law, in relation to the care, treatment and rehabilitation of narcotic addicts and making appropriations therefore. Dated 6 April 1966. (The "Rockefeller Law") (see ibid., 1969, 20, 166)

DRUG ABUSE: THE INTERNATIONAL POINT OF VIEW

V. Kusevic, M.D.

Director, U. N. Division of Narcotic Drugs, Geneva

**

A Meo tribesman in the mountains of Northern Thailand goes to a jungle clearing and lances the capsules of the opium poppy in his field, and an adolescent dies in New York of an overdose of heroin. A Lebanese farmer sells his crop of cannabis and young people in London, Paris or Stockholm get high and break their necks in a car crash. A chemist synthetizes LSD in a clandestine laboratory in California and a youngster jumps through the window in Geneva. Such events are closely linked and transcend national frontiers.

Drug abuse is an international problem and each year makes it more so. Up to 20 years ago, drug abuse was relatively easy to analyze: it was mainly limited to certain regions in Asia and South America with some spots in other regions. During the last 15 years or so, however, the development of drug abuse has been quite different. In countries where it was unknown, it has appeared; in countries where it touched only the fringes of society, it is now well settled even in well-to-do groups and among people to whom it would have been abhorrent earlier. The causes of this trend are very difficult to define, and this is not the place to go into them in detail. There is no doubt that the conditions created by the huge development of technology and of the modern "consumers' society" have laid the ground for the upsurge of drug abuse in western countries while in other countries the social tradition, for instance of opium smoking, generally based on poverty, has evolved into the modern form of heroin taking. Without question, the abuse of drugs in whatever form, has a social origin as well as it has dramatic social consequences.

I want to try to delineate international elements in the phenomenon:

Firstly, the main one is that "the world is getting smaller" and this is due to the enormous increase in the means of transportation and in the communication media. Migrations have increased in number and variety in a way which is hardly believable: first the tourists; each year in Europe alone, mass migrations from North to South and from South to North take place, involving perhaps as many as 40 to 50 million people. Obviously, boundaries are for all practical purposes knocked down by such an enormous amount of travellers. Among these people, a small proportion - but still a large number - know the "delights" of hashish for instance, and bring some back home.

Secondly, the intense industrialization of some countries created a movement of workers coming by hundreds of thousands from places where drugs such as hashish or opium are well known and generally available and cheap. Some of these workers started to bring these drugs back with them.

Finally, there is a movement which, however small, is still of importance, that is what has been called "the Route"; this is the strange migration of youngsters from North America and Western Europe, but also from other countries, toward such "paradises" as Kabul and Kathmandu. Under the delusion or pretext of finding the sources of ancient wisdom, thousands of youngsters break their ties with their family or society and tend to create mini-societies which glorify and almost deify opium, hashish and other psychotropic drugs. What they find is not the wisdom of Asia but the dregs of Asian civilizations; they look for freedom and reach the most debased kind of slavery. By so doing they constitute a very real danger to countless other adolescents by the glamour which somehow attaches itself in their eyes to what they believe to be a truly worthwhile form of existence.

The universal development of drug abuse has been called "an explosion" and this word seems rightly applied to this new phenomenon. Society, before this explosion, used to defend itself against drug abuse by two main devices: The control through international treaties and repression through enforcement activities. The international treaties, from 1909 at the time of the Shanghai Conference until 1971 when the Convention on Psychotropic Substances was adopted in Vienna, have certainly played a very important part in preventing the spread of drug abuse but it is unfortunately obvious that these efforts, however great and fruitful, were not sufficient. At least they succeeded in preventing drugs from the licit trade to be a source of abuse: by now there is hardly any link between the legal drugs and the illicit ones. Drug abuse feeds generally on illicit drugs.

As regards enforcement, by itself it cannot prevent drug abuse; no country has, as yet, succeeded in creating the sort of sanitary cordon which would be strong enough to prevent drugs from entering it; 10% at the most of drugs smuggled in a country are seized. If one considers the quantities reported as caught in the illicit traffic, one is staggered by the huge mass of drugs which reach their destination. Apart from shutting itself totally from the rest of the world which is an absurd concept, no country can by itself suppress the flow of illicit drugs entering its territory and therefore stamp out drug abuse.

Any focus of drug addiction anywhere in the world constitutes a danger for all other countries and if the number of heroin addicts multiplies, say in Thailand, this in itself is a danger not only as regards the spread of heroin abuse in neighbouring countries but also in far away parts of the world. This is because the presence of a large number of addicts creates a current traffic which very soon spreads to other markets and encourages illicit production.

In the last years a new concept has developed among international bodies dealing with the drug abuse problem: the concept of a world plan. For years in the Commission on Narcotic Drugs for instance an endless and inconclusive debate has been waged between those people saying: "If you cut the number of your addicts, our illicit producers will have no market and will stop production" and the other Party saying: "If you were to suppress your illicit production, our addicts would have no drugs". This controversy could obviously lead nowhere, and it appeared that only if and when the two main elements of drug abuse, namely supply and demand were tackled together could any progress be made. The third element, namely the illicit traffic is a secondary element which would die out slowly if the two main ones diminished, and which conversely cannot be fought against efficiently as long as these two elements flourish.

This is not the place to review in detail the question of the supply: suffice it to say that an evaluation has been made by the International Narcotics Control Board according to which more than 1200 tons of illicit or uncontrolled opium are produced in the world; there are also, the enormous quantities - which nobody can evaluate - of cannabis either growing wild or cultivated for illicit purposes. Finally, there is a huge production of coca leaf perhaps up to 30, 000 tons of which 98% is used for non-medical and non-scientific purposes.

The illicit and uncontrolled production of opium, cannabis and coca leaf is in practically all cases done by groups of population which dwell in the less-developed and poorest areas of a number of developing countries. The money that they earn from the sale of these products constitutes for them the main, if not the only cash income. The suppression of these crops by repressive means could only lead to social troubles and it is in fact impossible given the resources of the countries involved. The only way to suppress such cultivation is to give to the people engaged in it the means of earning as much through other occupations. This entails a change in the socio-economic conditions of the regions concerned and is a difficult and long lasting process.

As regards the demand, here again the repression by itself does not go to the root of the problem. Action against the demand must be mainly directed towards prevention, and secondly to the treatment, rehabilitation and social reintegration of addicts. Prevention has to be done through education of the public, especially youth. All means of treatment, rehabilitation and social reintegration of those who have been already contaminated must be explored through all avenues of research and action, mainly social so as to modify the socio-economic conditions which often constitute the ground for the growth of drug abuse.

Only when these two series of measures dealing with supply and demand are well under way and begin to bear fruit can the repression, that is mainly the fight against illicit traffic, obtain valid results.

These concepts may appear over ambitious; they have, however, to be translated into facts if the international community wishes to get rid or at least to reduce to insignificant proportions the problem of drug abuse. The international bodies concerned have finally defined a line of action which is summarized in the resolution adopted by the Commission on Narcotic Drugs at its second special session in October 1970; I shall quote part of that resolution which has afterwards been agreed to by the Economic and Social Council and the General Assembly of the United Nations:

"Integrated international action against drug abuse should be designed to:

(i) limit the supply of drugs to legitimate requirements by putting an end to their illegal or uncontrolled production, processing and manufacture, making use of crop substitution or other methods, as appropriate;

(ii) improve the administrative and technical capabilities of existing bodies concerned with the elimination of the illicit traffic in drugs;

(iii) develop measures to prevent drug abuse through programmes of education and special campaigns, including the use of mass media; and

(iv) provide facilities and develop methods for the treatment, rehabilitation and social reintegration of drug dependent persons"

To implement that resolution the Commission requested the Secretary-General of the United Nations to elaborate after consultation with UN bodies, the Specialized Agencies and other competent organizations, a plan for concerted action against drug abuse, including proposals for permanent arrangements regarding administration and financing, which would ensure proper co-ordination within the UN system. This plan has been drafted and examined by an inter-agency meeting which took place in July 1971. It describes a multidisciplinary action to be spread over a period of several years in which the United Nations, a number of its specialized agencies such as FAO, ILO, UNESCO, WHO, etc., and other bodies such as INCB, ICPO,

Interpol, etc., will participate.

The implementation of such a programme will obviously cost a large amount of money. For that reason the Commission recommended to the Economic and Social Council to invite the Secretary-General to set up a UN Fund for Drug Abuse Control to finance these activities. The Economic and Social Council agreed and invited the Secretary-General to establish the Fund, and the Secretary-General did so on 1 April 1971. This Fund is to be financed by voluntary contributions from governments and also private persons.

To conclude, the words used by the Secretary-General in the Aide-Memoire which was sent to governments prior to establishing the Fund might usefully be recalled here:-

> "In the past ten years, drug abuse and drug addiction have taken on increasingly dangerous proportions in many parts of the world. The situation is aggravated by the continuous introduction of new psychotropic substances which are liable to be misused. This trend has affected not only developed countries but also developing countries, and is especially dangerous in the latter because they have limited resources to deal with the situation. The reasons for the spread of drug abuse are complex and are different in different countries, and some of them are very difficult to remove; nevertheless, effective remedial action can and must be taken. It is clear that the problem requires international measures to deal with it; it is clear in broad outlines what those measures should be, and it is also clear that, unless corrective steps are taken, the present trend can lead only to disaster on an international scale. Concerted and simultaneous action must be taken on the supply of drugs for purposes of abuse, on the demand for such purposes and on the illicit traffic which serves as a channel connecting production with demand."

** Paper distributed but not read at Conference.

PARTICIPANTS

ANDREASSON, Dr. Rune, Executive Secretary General, International Association for Accident and Traffic Medicine, Karlavägen 119, S-115 26 Stockholm, Sweden

BASH, Prof. Kenower, Professor of Psychiatry, University of Berne, Psychiatrische Universitätsklinik, Bolligenstrasse 117, 3072 Ostermundigen-Berne, Switzerland

* BEEDLE, Mr. P., Home Office, Romney House, Marshal Street, London S.W. 1, U.K.

* BONNICHSEN, Prof. Roger, Institute for Forensic Chemical Laboratory, Tomtebodavägen, Stockholm 60, Sweden

CAMERON, Dr. D. C. Chief, Drug Dependence, WHO Geneva

CHRUSCIEL, Dr. T. L., Medical Officer, Drug Dependence, WHO, Geneva

DENIKER, Pierre, Professeur agrégé, Faculté de Médecine, 17 rue de la Ville l'Evêque, Paris VIIIe

DUBE, Kalika Charan, Professor of Psychiatry & Superintendent, Mental Hospital, Agra 2 U.P., India

EDWARDS, Dr. Griffith, Research Worker, Addiction Research Unit, Institute of Psychiatry, 101 Denmark Hill, London S.E. 5, U.K.

EVANG, Dr. Karl, Director-General of Health Services, Helsedirektoratet, The Royal Norwegian Ministry of Social Affairs, Oslo Dep., Norway

FREEDMAN, Dr. Daniel X., Louis Bloch Professor of Biological Sciences and Chairman, Department of Psychiatry, University of Chicago, Chicago, 60637, Illinois, U.S.A.

HUSAIN, Dr. Akhtar, Programme Specialist, Department of School and Higher Education, UNESCO, Place de Fontenoy, Paris VIIe

ISBELL, Dr. Harris, Professor of Medicine, University of Kentucky, Medical Center, Lexington 40505, Kentucky, U.S.A.

JACOB, Professeur Joseph, Jean, Chef du Service de Pharmacologie et de Toxicologie, Institut Pasteur, 26 rue du Docteur Roux, Paris XVe

KIELHOLZ, Prof. Paul, Dr. med., Director Psychiatrische Universitätsklinik Basel, Wilhem Klein St. 28, Basel, Switzerland

* KUSEVIC, Dr. V. Director, United Nations Division of Narcotic Drugs, Palais des Nations, Geneva, Switzerland

LEONG, Dr. Hon Koon, Medical Officer, Dept. of Social Medicine and Public Health, Faculty of Medicine, University of Singapore, Outram Hill, Singapore 3

MABILEAU, Dr. Jean, Inspecteur général, Inspection générale des Affaires sociales, Ministère de la Santé publique, 1, Place de Fontenoy, Paris VIIe

MENDELSON, Prof. Jack, Professor of Psychiatry, Harvard Medical School, Boston City Hospital, 818 Harrison Avenue, Boston, Mass. 02118 U.S.A.

MOERLOOSE, Dr. Jean de, Chief, Health Legislation, WHO Geneva

REXED, Prof. Bror, Director-General, National Board of Health and Welfare, Henriksdalsringen 61 IV, 13105 Nacka, Sweden

SHOHAM, Mr. Shlomo, Director, Institute of Criminology and Criminal Law, Tel-Aviv University, Tel Aviv, Israel

SMITH, Roger C. D. Crim., Executive Director, Marin Open House, 1327 Lincoln Avenue, San Rafael, California 94901, U.S.A.

STEINIG, Mr. Leon, Member of the International Narcotics Control Board, 1248 Hermance, Geneva, Switzerland

STEWART, Mr. F., Home Office, Romney House, Marsham Street, London S.W.1, U.K.

Chairman:

GELLHORN, Dr. Alfred, Dean, School of Medicine, University of Pennsylvania, Philadelphia, Pennsylvania, U.S.A. - President of CIOMS

Secretary:

BTESH, Dr. Simon, Executive Secretary, CIOMS

REPRESENTATIVES OF CIOMS MEMBER ORGANIZATIONS

World Federation of Societies of Anaesthesiologists Professeur Marcel GEMPERLE, Hôpital cantonal, Geneva

International Society of Criminology, Dr. Georges FULLY, Dr. Yves ROUMAJON, 5 rue Cambon, Paris 1er

International Dental Federation, Dr. Roland PETZOLD, Elisabethenanlage 7, CH-4051 Basel

International Epidemiological Association, Dr. Juraj CERVENKA Head, Dept. of Epidemiology, Postgraduate Medical School, (I.E.A. Secretary for Regional Development), Palisady 47, Bratislava, Czechoslovakia

International Society of Geographical Pathology, Prof. Jules ANGST, Dr. med., Director of Research, Prof. of Psychiatry, Psychiatrische Universitätsklinik,Lenggstr. 28, CH-8008 Zurich, Switzerland

International Union for Health Education, Dr. Jacques BASSOT, Secrétaire exécutif, 23 Rue Robert Doisy, 92-Antony, France

International Academy of Legal Medicine and Social Medicine
* Prof. Branko VOLARIC, Ul. Narodnog ustanka 9 a, 51000 Rijeka, Yugoslavia

International Society of Lymphology, Dr. Johannes BIRCHER, M.D., Oberarzt, Institut für Klinische Pharmakologie, University of Berne, Friedbühlstr. 49, 3008 Berne, Switzerland

World Medical Association, Dr. Anne AUDEOUD-NAVILLE, 10, Rue Bellot, Geneva 1206, Switzerland

World Federation for Mental Health, Dr. Anne AUDEOUD-NAVILLE, 10, rue Bellot, Geneva 1206, Switzerland

* International Congress of Neurology, Dr. J. M. ESPADALER, Palacio de las Naciones, Avda Maria Cristina, Parque de Montjuich, Barcelona, Spain

International Society of Neuropathology, Dr. Jean Emmanuel GRUNER, 90, rue de Varenne, Paris VIIe, Prof. G. MACCHI, Clinica delle Malattie Nervose e Mentali della Università di Perugia, via E. dal Pozzo, 95 - 06100 Perugia, Italy

International Paediatric Association, Prof. Ole WASZ-HÖCKERT, Prof. and Chairman, Dept. of Pediatrics, University of Oulu, Oulu, Finland

International Academy of Pathology, Dr. G. J. CUNNINGHAM, Virginia Commonwealth University, Health Sciences Division, Richmond, Va 23220, U.S.A.

World Association of Anatomic and Clinical Pathology Societies Dr. J. UNGAR, Serum und Impfinstitut, postfach 2707, 3001 Berne, Switzerland

World Psychiatric Association, Dr. Aldo CALANCA, Medical Chief, Psychiatric Clinic, University of Lausanne, Hôpital de Cery, 1008 Prilly, Switzerland

Medical Women's International Association, Dr. Renée Voluter de LORIOL, 196 route de Florissant, 1211 Conches-Switzerland

* International College of Surgeons , Prof. Alfonso de la FUENTE CHAOS, Catedra de Patologia Quirurgica, Hospital Clinico
* de la Facultad de Medicina, Madrid, Spain; Prof. Roelof RUDING, Haringvliestraat 6, Amsterdam, Netherlands

Royal Flemish Academy of Medicine, Belgium, Prof. André de SCHAEPDRYVER, Prof. of Pharmacology, University of Gent, School of Medicine, Heymans Institute of Pharmacology, De Pintelaan 135 B-9000, Gent, Belgium; Prof. Marc VERSTRAETE, Minderbroedersstraat 29, 3000 Leuven, Belgium

* Académie Royale de Médecine de Belgique, Prof. J. BOBON,
* Quai de Rome 14, 4000 Liège; Prof. P. DUMONT, Minckelersstraat 1, 3030 Heverlee, Belgium; Dr. Jean TITECA, Chargé du cours d'EEG à l'Université Libre de Bruxelles Médecin-chef du Centre Neuro-Psychiatrique de Bruxelles, 1er Vice-Président de l'Académie Royale de Médecine de Belgique, avenue de l'Exposition 341, B-1090 Bruxelles; Dr. Fernand GOFFIOUL, Maitre de Conférences à l'Université de Liège, Fédération des bureaux de consultations pour alcooliques et toxicomanes de Belgique, 11 avenue de Luxembourg, 4000 Liège

Czechoslovak Medical Society J.E. Purkyne - Czech Medical Association; Prof. Jaroslav HOREJSI, M.D., Charles' University, U nemocnice 1, Prague 2, Prof. Teofil Rudolf NIEDERLAND, M.D., D.Sc., President of Czechoslovak Medical Society J. E. Purkyne, 3-d Medical Clinic, Bratislava-Kramare

Royal Danish Academy of Sciences and Letters - Danish Medical Research Council, Prof. Villars LUNN, M.D., Solbakkevej 61, Gentofte, Denmark

Académie Nationale de Médecine (France), Prof. Jean CHEYMOL, 16 rue Bonaparte, Paris VIe

Federation of Hungarian Medical Societies, Hungarian Ministry for Health - Hungarian Committee against Alcoholism; Prof. Miklos KUN, Associate Professor of Psychiatry, University of Budapest, Chief Physician, Keleti-K-U 27 Budapest 11, Hungary

Israel Academy of Sciences and Humanities, Prof. Habib EDERY, Israel Institute for Biological Research, Tel-Aviv University Medical School, Ness Ziona, Israel

Centre Scientifique de Monaco, Dr. Robert SCARLOT, 1, Bd de Suisse, Monte-Carlo, Principauté de Monaco

Royal Netherlands Academy of Sciences and Letters, Prof. Joseph Jules Guillaume PRICK, Professor of Psychiatry and Neurology, Orangesingel 45, Nijmegen, Netherlands

South African Medical Research Council, Mr. O. A. van der WESTHUYSEN, Scientific Attaché, South African Embassy 38, rue de Bassano, Paris VIIIe

Swedish Medical Research Council, Prof. Bror REXED, Director-General, National Board of Health and Welfare, Henriksdalsringen 61 IV, 13105 Nacka, Sweden

Swiss Academy of Medicine, Dr. R. Dieter LADEWIG, Head, Addiction Research and Treatment Unit, Dept. of Psychiatry, University of Basel, Psychiatrische Universitäts-klinik, Wilhelm-Kleinstr. 27, Basel, Switzerland

National Academy of Sciences - National Research Council (U.S.A.), Dr. Daniel X. FREEDMAN, Louis Bloch Professor of Biological Sciences and Chairman, Department of Psychiatry, University of Chicago, Chicago 60637, Ill.

European Brain and Behaviour Society, Prof. Jacques PAILLARD, Directeur de Recherche, Professeur à l'Université Centre National de la Recherche Scientifique, Institut de Neurophysiologie et Psychophysiologie, 31 chemin Joseph Aiguier, 13 Marseille 9e, France

International Society of History of Medicine, Dr. Roger JORIS, 2 rue de la Porcelaine, 1260 Nyon, Switzerland

International Committee of Military Medicine and Pharmacy, Prof. Jovica PATRNOGIC, Professeur à la Faculté de Droit de Genève, 50, avenue d'Aïre, 1203 Geneva

International Association for Accident and Traffic Medicine, Dr. Rune ANDREASSON, Executive Secretary General, Karlavägen 119, S-115 26 Stockholm

REPRESENTATIVES OF UNITED NATIONS AND SPECIALIZED AGENCIES

United Nations Fund for Drug Abuse Control, Mr. C.W.A. SCHURMANN, Personal Representative of the Secretary General of the United Nations in charge of the U.N. Fund for Drug Abuse Control, Palais des Nations, Geneva, Switzerland

United Nations Commission on Narcotic Drugs, Mr. V.
* KUSEVIC, Director, U.N. Division of Narcotic Drugs, Geneva

* International Narcotics Control Board, Mr. J. DITTERT, Secretary of the Board, Geneva; Dr. N. BARCOV, Senior Scientist, Profsojusnaja st. 102, kor 5, kv 160, Moscow 117485 Mr. L. MANUECO-JENKINS, Legal Analyst (Social Affairs Officer) Palais des Nations, Geneva

United Nations Educational, Scientific and Cultural Organization Dr. Akhtar HUSAIN, Programme Specialist, Department of School and Higher Education, Place de Fontenoy, Paris VIIe

World Health Organization, Dr. T. A. LAMBO, Assistant Director General; Dr. V. FATTORUSSO, Director, Division of Pharmacology and Toxicology; Dr. D. C. CAMERON, Chief, Drug Dependence; Dr. T.L. CHRUSCIEL, Medical Officer, Drug Dependence; Dr. J. de MOERLOOSE, Chief, Health Legislation; Dr. R. ELLING, Chief, Behavioural Sciences

OBSERVERS

Council of Europe, Mr. Hans PFEFFERMANN, Head, Public Health Division, Place Lenôtre, 67 Strasbourg

International Council on Alcohol and Addictions, Mr. Archer TONGUE, Executive Director, POB 140, 1001 Lausanne, Switzerland

International Federation of Pharmaceutical Manufacturers Association, Dr. Walter P. von WARTBURG, 123 Grenzacherstrasse, Bâle, Switzerland; Dr. Paolo PRIORE, Assofarma, Piazza Velasca 5, Milan, Italy

Bundesverband der Pharmazeutischen Industrie e.V. Dr. Hanns-Reinhold BIENERT, Bingerst. 11, 6 Mainz, G Germany

International College of Psychosomatic Medicine, Prof. Jules ANGST, Dr. med., Director of Research, Prof. of Psychiatry, Psychiatrische Universitätsklinik, Lenggstr. 28, CH-8008 Zurich, Switzerland

League of the Red Cross Societies, Dr. Henryk ZIELINSKI, Assistant Director, Health and Social Service Bureau, Dr. Vladimir SEMUKHA, Expert-consultant on Health Problems 17, chemin des Crêts, Petit-Saconnex, 1211 Geneva 19

Austrian Academy of Sciences, Prof. Heribert KONZETT, Reithmannstr. 18, A 6020 Innsbruck, Austria

Department of Medicine, College of Medicine, National Taiwan University, Prof. Wan-Yu CHEN, No. 126 Tatong Road, Shih Lin, Taipei, Taiwan

East African Medical Research Council of the East African Community, Dr. Fabian KAMUNVI, Secretary, P.O.B. 1002, Arusha, Tanzania, East Africa; Dr. Hamza Juma CHUM, Medical Research Officer, E.A. Leprosy Research Center, P.O.B. 1044, Busia, East Africa

Centre National de la Recherche Scientifique (France), Prof. Jacques PAILLARD, Directeur de Recherches, Prof. à l'Université, Dr. Robert NAQUET, Directeur de Recherches Institut de Neurophysiologie et de Psychophysiologie, 31 chemin Joseph Aiguier, 13 Marseille 9e

* Centre de Toxicologie (France), Prof. GAUTTIER, Hôpital Fernand Widal, 200 rue du Faubourg St. Denis, Paris

Institut de France - Académie des Sciences, Prof. René TRUHAUT, Professeur de Toxicologie à la Faculté des Sciences pharmaceutiques et biologiques de l'Université René Descartes, Paris Ve - Faculté de Pharmacie, 4 avenue de l'Observatoire, Paris VIe

Institut National de la Santé et de la Recherche Médicale (France), Prof. Agrégé Pierre DENIKER, Faculté de Médécine, 17 rue de la Ville l'Evêque, Paris VIIIe , Dr. Jean Emmanuel GRUNER, Directeur de Recherches, 90 , rue de Varenne , Paris VIIe

Institut Pasteur (France), Prof. Joseph Jean JACOB, Chef du Service de Pharmacologie et de Toxicologie, 26 rue du Docteur Roux, Paris XVe

Max-Planck Institut for Psychiatry (Max-Planck Society for the Advancement of Sciences) (Germany) Prof. Priv. Doz. Wilhelm FEUERLEIN, Kraepelinstrasse 10, D8 Munich 23

Conseil National des Recherches Scientifiques et Techniques (Haiti), Prof. Victor NOEL M.D. Ph.D., President du Conseil National des Recherches Scientifiques et Techniques, 116 rue Capois, Port-au-Prince, Haiti

Istituto Superiore di Sanità (Institute of Health) - Italy,
* Prof. BIGNAMI, Viale Regina Elena 299, Rome

Lebanese Council for Scientific Research - Lebanon, Dr. Abdul-Rahman LABBAN, Vice-President, Bir-Hassan, Beirut

Norwegian Council for Research in Science and Humanities, Prof. Nils RETTERSTOL, Prof. of Psychiatry, Neevengarden Hospital, University of Bergen, 5000 Bergen, Norway

* Pakistan Medical Research Council, Professor Mohammad IBRAHIM, S.K., Chairman, Jinnah Postgraduate Medical Centre, Karachi-35, Pakistan

* National Institute of Health (Peru), Dr. German NORIEGA, Apartado No. 451, Lima, Peru

Fonds National Suisse de la Recherche Scientifique (Swiss National Foundation for Scientific Research), Prof. Meinrad E. SCHAR, Professor of Social and Preventive Medicine, Institute of Social Medicine, 32 Gloriastrasse, 8006 Zurich, Switzerland

Permanent Mission of Thailand, Mr. Prasert MASOODI, First-Secretary, Royal Thai Embassy, 60 Eigerstrasse, 3000 Berne, Switzerland

Scientific and Technical Research Council of Turkey, Dr. Resat GARAN, Hüsrev Gerede Cad. 75/1, Tesvikiye, Istanbul

Medical Research Council (United Kingdom), Dr. Joan Mary FAULKNER, Principal Medical Officer, 20 Park Crescent, London, W.1

The Royal Society (United Kingdom), Dr. William PATON, Professor of Pharmacology, University Department of Pharmacology, South Park Rd, Oxford, U.K.

Social Science Research Council (United Kingdom), Dr. Simon YASIN, Principal Scientific Officer, State House, High Holborn, London W.C.1

National Institute of Mental Health, Department of Health, Education and Welfare - U.S.A. Government, Dr. Lorrin M. KORAN, Special Assistant to the Director, 5600 Fisher's Lane, Room 17C17, Rockville, Maryland 20852, U.S.A.

* Dr. W. BUNNEY, Director, Division of Narcotic Drugs, U.S.A.

Permanent Vietnamese Mission to the International Organizations at Geneva, Mr. Pham van TRINH, Second Secretary, Miss Do Thi Thuy NGOÇ, Third Secretary, 58, rue de Moillebeau, 1211 Geneva 9

Laboratoires Winthrop (France), Dr. Michel AURICHE Directeur de la Recherche Médicale, Dr. Anne-Catherine ARDOUIN, Directeur du Service des Nouveaux Produits, 92 - 98 Boulevard Victor Hugo, 92 - Clichy, France

Winthrop Laboratories - (United Kingdom), Dr. John Michael MUNGAVIN, Medical Director, Dr. John B. SPOONER, Associate Medical Director, Surbiton-upon-Thames, Surrey, United Kingdom

* Unable to come